Characteristics and Uses of Steel Slag in Building Construction

Related titles

Nonconventional and Vernacular Construction Materials

(ISBN 978-0-08-100871-3)

Eco-efficient Materials for Mitigating Building Cooling Needs

(ISBN 978-1-78242-380-5)

Eco-efficient Masonry Bricks and Blocks

(ISBN 978-1-78242-305-8)

Woodhead Publishing Series in Civil and Structural Engineering: Number 67

Characteristics and Uses of Steel Slag in Building Construction

Ivanka Netinger Grubeša,
Ivana Barišić,
Aleksandra Fucic and
Samitinjay S. Bansode

ELSEVIER

AMSTERDAM • BOSTON • CAMBRIDGE • HEIDELBERG
LONDON • NEW YORK • OXFORD • PARIS • SAN DIEGO
SAN FRANCISCO • SINGAPORE • SYDNEY • TOKYO
Woodhead Publishing is an imprint of Elsevier

WP WOODHEAD PUBLISHING

Woodhead Publishing is an imprint of Elsevier
The Officers' Mess Business Centre, Royston Road, Duxford, CB22 4QH, UK
50 Hampshire Street, 5th Floor, Cambridge, MA 02139, USA
The Boulevard, Langford Lane, Kidlington, OX5 1GB, UK

Notices
Knowledge and best practice in this field are constantly changing. As new research and experience broaden our understanding, changes in research methods, professional practices, or medical treatment may become necessary.

Practitioners and researchers must always rely on their own experience and knowledge in evaluating and using any information, methods, compounds, or experiments described herein. In using such information or methods they should be mindful of their own safety and the safety of others, including parties for whom they have a professional responsibility.

To the fullest extent of the law, neither the Publisher nor the authors, contributors, or editors, assume any liability for any injury and/or damage to persons or property as a matter of products liability, negligence or otherwise, or from any use or operation of any methods, products, instructions, or ideas contained in the material herein.

British Library Cataloguing-in-Publication Data
A catalogue record for this book is available from the British Library

Library of Congress Cataloging-in-Publication Data
A catalog record for this book is available from the Library of Congress

ISBN: 978-0-08-100976-5 (print)
ISBN: 978-0-08-100376-3 (online)

For information on all Woodhead Publishing publications
visit our website at https://www.elsevier.com/

Publisher: Matthew Deans
Acquisition Editor: Gwen Jones
Editorial Project Manager: Charlotte Cockle
Production Project Manager: Debasish Ghosh
Designer: Greg Harris

Contents

Woodhead Publishing Series in Civil and Structural Engineering

1 **Finite element techniques in structural mechanics**
C. T. F. Ross
2 **Finite element programs in structural engineering and continuum mechanics**
C. T. F. Ross
3 **Macro-engineering**
F. P. Davidson, E. G. Frankl and C. L. Meador
4 **Macro-engineering and the earth**
U. W. Kitzinger and E. G. Frankel
5 **Strengthening of reinforced concrete structures**
Edited by L. C. Hollaway and M. Leeming
6 **Analysis of engineering structures**
B. Bedenik and C. B. Besant
7 **Mechanics of solids**
C. T. F. Ross
8 **Plasticity for engineers**
C. R. Calladine
9 **Elastic beams and frames**
J. D. Renton
10 **Introduction to structures**
W. R. Spillers
11 **Applied elasticity**
J. D. Renton
12 **Durability of engineering structures**
J. Bijen
13 **Advanced polymer composites for structural applications in construction**
Edited by L. C. Hollaway
14 **Corrosion in reinforced concrete structures**
Edited by H. Böhni
15 **The deformation and processing of structural materials**
Edited by Z. X. Guo
16 **Inspection and monitoring techniques for bridges and civil structures**
Edited by G. Fu
17 **Advanced civil infrastructure materials**
Edited by H. Wu
18 **Analysis and design of plated structures Volume 1: Stability**
Edited by E. Shanmugam and C. M. Wang
19 **Analysis and design of plated structures Volume 2: Dynamics**
Edited by E. Shanmugam and C. M. Wang

20 **Multiscale materials modelling**
Edited by Z. X. Guo
21 **Durability of concrete and cement composites**
Edited by C. L. Page and M. M. Page
22 **Durability of composites for civil structural applications**
Edited by V. M. Karbhari
23 **Design and optimization of metal structures**
J. Farkas and K. Jarmai
24 **Developments in the formulation and reinforcement of concrete**
Edited by S. Mindess
25 **Strengthening and rehabilitation of civil infrastructures using fibre-reinforced polymer (FRP) composites**
Edited by L. C. Hollaway and J. C. Teng
26 **Condition assessment of aged structures**
Edited by J. K. Paik and R. M. Melchers
27 **Sustainability of construction materials**
J. Khatib
28 **Structural dynamics of earthquake engineering**
S. Rajasekaran
29 **Geopolymers: Structures, processing, properties and industrial applications**
Edited by J. L. Provis and J. S. J. van Deventer
30 **Structural health monitoring of civil infrastructure systems**
Edited by V. M. Karbhari and F. Ansari
31 **Architectural glass to resist seismic and extreme climatic events**
Edited by R. A. Behr
32 **Failure, distress and repair of concrete structures**
Edited by N. Delatte
33 **Blast protection of civil infrastructures and vehicles using composites**
Edited by N. Uddin
34 **Non-destructive evaluation of reinforced concrete structures Volume 1: Deterioration processes**
Edited by C. Maierhofer, H.-W. Reinhardt and G. Dobmann
35 **Non-destructive evaluation of reinforced concrete structures Volume 2: Non-destructive testing methods**
Edited by C. Maierhofer, H.-W. Reinhardt and G. Dobmann
36 **Service life estimation and extension of civil engineering structures**
Edited by V. M. Karbhari and L. S. Lee
37 **Building decorative materials**
Edited by Y. Li and S. Ren
38 **Building materials in civil engineering**
Edited by H. Zhang
39 **Polymer modified bitumen**
Edited by T. McNally
40 **Understanding the rheology of concrete**
Edited by N. Roussel
41 **Toxicity of building materials**
Edited by F. Pacheco-Torgal, S. Jalali and A. Fucic
42 **Eco-efficient concrete**
Edited by F. Pacheco-Torgal, S. Jalali, J. Labrincha and V. M. John
43 **Nanotechnology in eco-efficient construction**
Edited by F. Pacheco-Torgal, M. V. Diamanti, A. Nazari and C. Goran-Granqvist

44 **Handbook of seismic risk analysis and management of civil infrastructure systems**
Edited by F. Tesfamariam and K. Goda
45 **Developments in fiber-reinforced polymer (FRP) composites for civil engineering**
Edited by N. Uddin
46 **Advanced fibre-reinforced polymer (FRP) composites for structural applications**
Edited by J. Bai
47 **Handbook of recycled concrete and demolition waste**
Edited by F. Pacheco-Torgal, V. W. Y. Tam, J. A. Labrincha, Y. Ding and J. de Brito
48 **Understanding the tensile properties of concrete**
Edited by J. Weerheijm
49 **Eco-efficient construction and building materials: Life cycle assessment (LCA), eco-labelling and case studies**
Edited by F. Pacheco-Torgal, L. F. Cabeza, J. Labrincha and A. de Magalhães
50 **Advanced composites in bridge construction and repair**
Edited by Y. J. Kim
51 **Rehabilitation of metallic civil infrastructure using fiber-reinforced polymer (FRP) composites**
Edited by V. Karbhari
52 **Rehabilitation of pipelines using fiber-reinforced polymer (FRP) composites**
Edited by V. Karbhari
53 **Transport properties of concrete: Measurement and applications**
P. A. Claisse
54 **Handbook of alkali-activated cements, mortars and concretes**
F. Pacheco-Torgal, J. A. Labrincha, C. Leonelli, A. Palomo and P. Chindaprasirt
55 **Eco-efficient masonry bricks and blocks: Design, properties and durability**
F. Pacheco-Torgal, P. B. Lourenço, J. A. Labrincha, S. Kumar and P. Chindaprasirt
56 **Advances in asphalt materials: Road and pavement construction**
Edited by S.-C. Huang and H. Di Benedetto
57 **Acoustic emission (AE) and related non-destructive evaluation (NDE) techniques in the fracture mechanics of concrete: Fundamentals and applications**
Edited by M. Ohtsu
58 **Nonconventional and vernacular construction materials: Characterisation, properties and applications**
Edited by K. A. Harries and B. Sharma
59 **Science and technology of concrete admixtures**
Edited by P.-C. Aïtcin and R. J. Flatt
60 **Textile fibre composites in civil engineering**
Edited by T. Triantafillou
61 **Corrosion of steel in concrete structures**
Edited by A. Poursaee
62 **Innovative developments of advanced multifunctional nanocomposites in civil and structural engineering**
Edited by K. J. Loh and S. Nagarajaiah
63 **Biopolymers and biotech admixtures for eco-efficient construction materials**
Edited by F. Pacheco-Torgal, V. Ivanov, N. Karak and H. Jonkers
64 **Marine concrete structures: Design, durability and performance**
Edited by M. Alexander
65 **Recent trends in cold-formed steel construction**
Edited by C. Yu
66 **Start-up creation: The smart eco-efficient built environment**
Edited by F. Pacheco-Torgal, E. Rasmussen, C. G. Granqvist, V. Ivanov, A. Kaklauskas and S. Makonin

About the authors

The authors of this book, I. Netinger Grubeša and I. Barišić, are civil engineers with many years of experience in researching slag utilization as building material. They are working at the Faculty of Civil Engineering Osijek, University of Osijek, teaching building materials and road building, respectively. In their scientific work, they are focused mainly on the application of all kind of waste materials in construction of civil engineering structures. Altogether they have published over 80 scientific papers, four books and three book chapters. Samitinjay S. Bansode, a civil engineer, also having many years of research experience in the field of Geo-Environmental Engineering, contributed to this book by giving insight into the range of impacts that steel slag could have in the construction industry. Bansode gave added value to this book by providing the considerable experiences of India in the disposal of this by-product. They were joined in this endeavor by Aleksandra Fucic, a genotoxicologist who contributed in data collection on the possible health or environmental effects caused by reutilizing slag in buildings, thus ensuring an interdisciplinary approach. She is expert in biomonitoring. During the last 30 years her main scientific interest are carcinogenesis mechanisms in subjects exposed to chemical and physical agents. She has published over 80 original papers and several books. She is teaching genotoxicology at Postgraduate studies at Medical School University of Zagreb.

Foreword

The construction sector is one of the most influential industries in terms of the environment, with a strong impact on waste production and energy consumption, as well as great potential for using waste products. The global economic crisis and European zero waste politics in recent years have promoted a more comprehensive utilization of waste and industrial by-products such as fly ash, construction waste, and slag in the construction sector.

On the other hand, the construction sector also consumes large quantities of natural materials, which calls for solutions that can reduce the related adverse environmental impacts. In addition, the technologies for exploiting natural materials cause various negative effects, including visual blight on the environment, increased heavy traffic on roads that cannot handle them well, noise, dust, and vibration. Therefore, in addition to the introduction of new solutions that would rationalize the usage of natural materials, it is crucial to enforce the production of construction materials from waste, thus reducing the cost of building and the size of dumping sites. Such an approach has been the incentive for researchers to focus on finding new methods in civil engineering to produce environmentally friendly structures.

Reflecting this trend, the primary aim of this book is to present all the many possibilities of steel slag for use as a building material and evaluate its properties before it is effectively incorporated into the corpus of standard construction materials and approved for regular usage. We are witnesses to the fact that, in the history of human technologies, many materials were abandoned after their shortcomings or related health risks were discovered. This book makes a contribution based on scientific investigations and an open-minded interdisciplinary approach in order to inform readers and motivate new investigations.

Steel slag, with its physical properties and controllable impact on the environment, has great potential to be included in the inventory of waste applied as construction material. This book has been prepared on the basis of scientific projects and the long-standing experience of its coauthors in the evaluation of the profile of steel slag as a by-product. It relies on investigations of best practices for its application following the dynamics of its production and its distribution in the global market.

During the period between 2008 and 2011, the possibilities of utilizing steel slag as a concrete aggregate were researched within the project "E!4166—EUREKABUILD FIRECON; Fire-Resistant Concrete Made with Slag from the Steel Industry." The properties of steel slag locally produced in Croatia were explored within the framework of this project, as were the properties of fresh and hardened concrete containing steel slag aggregate, observed under regular environmental exposure and fire exposure conditions. The Faculty of Civil Engineering in Zagreb coordinated the project, while the Faculty of Civil Engineering in Osijek and the Slovenian National Building and Civil Engineering Institute were partners. For the purposes of this project, coarse slag fractions were used as an aggregate for concrete production, and fine slag fractions proved to be a useful material that can be implemented in road construction. Extended research incorporated investigations into the properties of utilizing fine slag fractions in road construction. The entire corpus of the aforementioned project, as well as an abundant fund of photographs collected during research, has been provided in this book for the first time. The data presented form a core of knowledge regarding the utilization of slag that can be useful to civil engineers, as well as those with roles in waste management and environmental health.

Introduction

1

Civil engineering is an activity that essentially relies on exploiting natural resources. However, the ever-growing demand for materials by the building industry cannot be fully met by natural resources or traditional materials. Hence, there is a need to develop potential alternative materials and innovative techniques to solve the increasing demands of building construction. The response to this issue can be found in the reuse of waste materials. Furthermore, a large amount of waste results from the demolition caused during construction, and all of this has to be managed or disposed of somehow. The building material industry here comes to the fore as a domain of interest for reusing the waste material.

Even though waste materials are increasing today during the construction of new buildings and the rehabilitation of existing structures, civil engineering has left a very large ecological footprint throughout history. The influence is evident from the example of a 1-km-long, four-lane highway made of concrete pavement. This road requires about 1620 tons of cement, 7800 tons of coarse aggregate, and about 3240 tons of sand. If the same road were made of asphalt, it would require about 3600 tons of coarse aggregate, 2400 tons of fine aggregate, 540 tons of sand, and 300 tons of bitumen [1]. During aggregate preparation and other paving work, 1200 tons of CO_2 is produced, which is almost equal to the total CO_2 emissions produced by 210 passenger cars in a year [2]. Since the network of roads throughout the entire world is 15.99 million km long (for comparison, the distance between the Moon and the Earth is only 384,400 km), the implications of this statistic lead to alarming findings about the scale of the adverse environmental impact of road construction, as only one branch of civil engineering.

Water is the most consumed material in construction, but the runner-up is concrete. It is estimated that roughly 25 billion tons of concrete are manufactured globally each year, which amounts to more than 3.8 tons per person in the world [3]. It is mostly used in buildings, but it is also present in pavement. Besides the huge amount of used aggregate, due to the wide use of these materials, the cement and concrete industries are the biggest CO_2 producers, with cement production contributing about 5% of annual anthropogenic global CO_2 production [4]. Therefore, in recent years, researchers have focused on finding new methods of design, construction, and maintenance with the purpose of producing environmentally friendly buildings. Most of these studies are based on the use of waste materials, which solves the problem of waste disposal but also contributes to the savings and preservation of natural, nonrenewable materials. Therefore, the civil engineering community, aware of this negative trend, is turning to exploring the ecological principles of building, primarily through using lesser amounts of natural, nonrenewable resources. At the same time, we must face the pressing problem of disposing of the increasing amount of various wastes.

Characteristics and Uses of Steel Slag in Building Construction. http://dx.doi.org/10.1016/B978-0-08-100368-8.00001-4

1.1 Legal framework for waste management

Today, waste is one of the key problems faced by the world in general, and civil engineering has been trying to address this issue by following the principles of sustainable development. As materials from natural resources are usually either already present on a construction site or are brought there from a nearby site, most of the standards for civil engineering materials are based on the assumption that natural materials are being used on a building project. In order to ensure the transfer of knowledge about waste materials obtained in research to practice in real life, a legal framework is needed.

One of the first legal documents pertaining to environmental preservation was the Basel Convention on the Control of Transboundary Movements of Hazardous Wastes and Their Disposal, passed in 1989. Another very important international document was the Kyoto Protocol, which was adopted in Kyoto, Japan, in 1997 and came into force in 2005. This document is a supplement to the already existing United Nations Framework Convention on Climate Change, and it was signed with the aim of reducing greenhouse gas emissions. The states that have ratified it create 61% of the world's pollutants. Today, all European Union (EU) member states must adapt their laws to the current Waste Framework Directive (WFD), which provides a legislative framework for the collection, transport, recovery, and disposal of waste and includes a common definition of waste [5]. The revised WFD came into force in 2008, and its requirements are supplemented by other directives for specific waste streams. The directive also requires EU member states to take appropriate measures to encourage (i) the prevention or reduction of waste production and (ii) the recovery of waste by means of recycling, reuse, reclamation, or any other process, with a view to extracting secondary raw materials or using the waste as an energy source. Prior to this document, there was no definition of the term *by-product* in legislation in any European country. This document clearly defines "*by-product* as a substance or object, resulting from a production process, the primary aim of which is not the production of that item". By-products, therefore, are production residue, not waste. Material can be considered as by-product if it meets all of the following criteria [6]:

- Further use of the substance or object is certain.
- The substance or object can be used directly, without any further processing other than normal industrial practice.
- The substance or object is produced as an integral part of a production process.
- Further use is lawful; i.e., the substance or object fulfils all relevant product, environmental, and health protection requirements for the specific use in question and will not lead to overall adverse environmental or human health impacts.

According to the directive, *waste* is defined as "any substance or object which the holder discards or intends or is required to discard". Despite this given definition, many publications still use the term *waste* to refer to alternative materials that can be used in building rather than materials from natural resources.

1.2 Alternative materials in civil engineering

In civil engineering, the term *alternative materials* usually refers to solid wastes generated by industrial, mining, domestic, and agricultural activity. The type and nature of solid wastes and their recycling, as well as their utilisation potential in civil engineering, are listed in Table 1.1.

When solid waste is used in place of other conventional materials, natural resources and energy are preserved and expensive and potentially harmful waste disposal methods are avoided. Other advantages of using waste include reduced energy consumption using already existing materials, reduced pollution and global warming, and reduced waste in landfills. However, using waste materials is not always cost effective, because setting up new recycling units can be a high upfront cost.

Table 1.1 Types and nature of solid wastes and their recycling and utilisation potential [7–9]

Type of solid waste	Source details	Recycling and utilisation in building applications
Agro waste (organic)	Baggage, rice and wheat straw and husk, cotton stalks, saw mill waste, ground nut shells, banana stalks, and jute, sisal, and vegetable residue	Particleboard, insulation boards, wall panels, printing paper and corrugating media, roofing sheets, fuel, binders, fibrous building panels, bricks, acid-proof cement, coir fibre, reinforced composite, polymer composites, cement board
Industrial waste (inorganic)	Coal combustion residue, slag, bauxite red mud, waste glass, rubber tires, construction debris	Cement, bricks, blocks, tiles, paint, aggregate, cement, concrete, wood substitute products, ceramic products, subbase pavement materials
Mining/mineral waste	Coal washery waste, mining overburden waste, quarry dust, tailing from the iron, copper, zinc, gold, aluminium industries	Bricks, tiles, aggregates, concrete, surface finishing materials, fuel
Nonhazardous other process waste	Waste gypsum, lime sludge, limestone waste, marble processing residue, broken glass and ceramics, kiln dust	Gypsum plaster, fibrous gypsum board, bricks, blocks, cement clinkers, supersulphate cement, hydraulic binders
Hazardous waste	Metallurgical residue, galvanising waste, tannery waste	Cement, bricks, tiles, ceramics, and boards

1.3 Slag as an alternative building material

Slag is a broad term covering all nonmetallic coproducts resulting from the separation of a metal from its ore. Its chemistry and morphology depend on the metal being produced and the solidification process used [10]. Slag, as a material, is as old as the melting process in which it is produced. The various metal melting processes, the types and properties of slag generated (depending on the melting process), and the history of slag utilisation are described in the following sections.

1.3.1 Metal melting processes

The first melting process consisted of heating crushed ore and coal in a clay furnace, whose temperature was increased by blowing air through a clay pipe [11]. In this process, carbon separated the metal from its oxide and carbonate and evaporated in the form of CO_2, and eventually slag and pure metals (e.g., copper) precipitated at the bottom of the furnace due to the higher density of those materials. After processing, metal was poured into molds made of stone or fired clay. In time, these furnaces become more complex.

While the earliest records of metal melting in West Asia date back to 5500–5000 BC, iron was not melted before 2000 BC [11]. In ancient times, only the Chinese were producing cast iron (and thus slag), since they were able to achieve the required temperature for melting the iron by constructing a high-quality blast furnace from refractory clay. However, significant advances in the technology of casting iron occurred in the 18th century, when Quaker ironmaster Abraham Darby constructed blast furnace coal as a fuel [11].

A blast furnace is a huge, steel stack lined with refractory brick, where iron ore, coke, and limestone are dumped into the top, and preheated air is blown into the bottom. The first blast furnaces were built in the 14th century and produced 1 ton of iron per day, and even though equipment improved and higher production rates could be achieved (up to 13,000 tons of iron per day), the processes inside the blast furnace remained the same [12]. The first true blast furnace (i.e., a furnace with the ability to produce fluid crude iron) included all devices exceeding 12 feet (3.7 m) in height [13].

The actual geographical origin of the blast furnace is unclear. A cast-iron faun figurine was found in the Athens art trade in 1907, as well as a vase with a picture of a blast furnace. The ancient Mumbwa tribe of Africa consistently produced cast iron with a bone and quartz flux. Large cast-iron ingots of the Roman period have been found in France. However, there is a consensus that only in ancient China and early modern Europe did the primeval cast-iron-producing furnace become popularly used.

The work principle and purpose of a blast furnace [i.e., to chemically reduce and physically convert iron oxides into liquid iron ("hot metal")] are the same even today. Charges are heated and dried by hot gases that rise from below as they fall into a fiery bed of glowing coals at the bosh (about the midpoint of the furnace, where chemical reactions begin). Meanwhile, a blast of cold air is forced into the bosh from below, furnishing oxygen to intensify the heat and help keep the materials from falling to the bottom of the furnace [14]. At high temperatures, iron ore transforms into iron and carbon dioxide or carbon monoxide with melted iron, as heavy, molten fluid sinks to the crucible or bottom of the furnace. Meanwhile, limestone becomes a fluxing material, uniting with other impurities in the ore to form a molten waste fluid that sinks into the crucible. Liquid slag then trickles through the coke bed to the bottom of the furnace, where it floats on top of the liquid iron since it is less dense. The texture and color of slag indicate which ore is used. Dark gray slag indicates a high grade of ore, green or black indicates a protoxide ore, brown comes from magnetic ore, dirty yellow or red comes from peroxide ore, and turquoise blue is seen when manganese is in the ore [14].

Besides slag, hot, dirty gases exit the top of the blast furnace and proceed through gas cleaning equipment. Particulate matter is removed from the gas, the gas is cooled, and due to its considerable energy, it is burned as a fuel in hot blast stoves, which are used to preheat the air entering the blast furnace. Any gas not burned in the stoves is sent to the boiler house and is used to generate steam, which turns a turbo blower that generates compressed air (known as *cold blast*) that comes to the stoves [12].

Steelmaking has played a crucial role in the development of modern technological societies. Iron is a hard, brittle material that is difficult to work with, whereas steel is relatively easily formed and versatile. The mass production of steel started with the invention of the Bessemer converter in the late 1850s. The key principle in use in that device was the removal of impurities from the iron by oxidation, with air being blown through molten iron. At the time, the Thomas converter was in use too. The difference between the two types of converters was the main source of heat; in the Bessemer converter, it is silicon, while in the Thomas converter, it is phosphorus, whose content in pig iron may be as high as 2%. A more refined version of the Bessemer converter, where the blowing of air was replaced with blowing oxygen was commercialized in 1952–1953. The process that takes place in such a furnace is known as the *Linz-Donawitz process* or *basic oxygen process,* and this refined these furnaces into basic oxygen furnaces (BOFs).

Open-hearth furnaces were first developed by the German engineer Carl Wilhelm Siemens. In 1865, the French engineer Pierre-Émile Martin took out a license from Siemens and first applied his regenerative furnace for making steel. Their process was known as the *Siemens-Martin process,* and the furnace was known as an *open-hearth* or *Siemens-Martin (SM) furnace.* The open-hearth furnace is charged with light scrap, such as sheet metal, shredded vehicles, or waste metal, and heated using burning gas. Once it has melted, heavy scrap, such as building, construction, or steel milling scrap, is added, together with pig iron from blast furnaces. Once all the steel has melted, slag-forming agents such as limestone are added. The oxygen in iron oxide and other impurities decarburize the pig iron by burning excess carbon away, forming steel. To increase the oxygen content of the heat, iron ore can be added. Most open-hearth furnaces were closed by the early 1990s, to be replaced by the BOF or electric arc furnace (EAF).

These days, steel is widely produced by using electric power in EAFs. Comparing to blast furnaces, the history of EAFs is quite short. The EAF applied in steelmaking was invented in 1889 by the French scientist Paul Héroult utilising electric energy, which was relatively cheap at that time [15]. The first-generation furnaces had a capacity of between 1 and 15 t. Its main advantage over other steelmaking devices (such as Bessemer converters and open-hearth furnaces) was the possibility of producing special steels requiring high temperatures, ferroalloy melting, and long refining times. Today, EAF produces 29% of the crude steel produced worldwide, while China, the United States, and India are the world leaders in EAF production [15].

The only obstacle encountered when producing certain specific steel grades in EAFs is the contamination of scrap with copper, nickel, chrome, and other residual contaminants, which cannot be removed in the course of processing the finished steel. Permissible content of these contaminants is strictly limited in high-quality steel grades [16]. The EAF operating cycle is comprised of six operations (furnace charging, melting, refining, deslagging, tapping, and furnace turnaround) and today's modern operations aim for a tap-to-tap time of less than 60 min [17]. With furnace charging, it is important to select which grade of steel to make to ensure proper melt-in chemistry and melting conditions. The melting period, at a heart of EAF operations, is accomplished by supplying electrical or chemical energy to the furnace interior. Electrical energy is supplied via graphite electrodes and is usually the largest contributor in melting operations, while chemical energy is supplied via several sources, including oxy-fuel burners and oxygen lances. During this phase, dust is formed that contains mainly iron oxides, CaO, and ZnO [15]. This dust is typically collected to bag filters where it can be recycled in the EAF itself, reducing total dust generation per year and per ton of produced steel. The Zn content is increasing cycle by cycle, and the dust removed from the circuit has 20% ZnO or more, making it more attractive to zinc producers. Refining operations involve the removal of not only phosphorus, sulphur, aluminium, silicon, manganese, and carbon from the steel, but also dissolved gases, especially hydrogen and nitrogen [17]. In deslagging operations, impurities are removed from the furnace with the furnace tilted backward and slag pours out through the slag door. When the desired steel composition and temperature in the furnace are achieved, the tapping process begins, during which bulk alloy additions are made based on the bath analysis and the desired steel grade. Finally, during the last phase of the EAF process, furnace turnaround is conducted, which is the period following the tapping until the furnace is recharged for the next heating.

1.3.2 Slag types

Slag, the product generated by the purification, casting, and alloying of metals, is also classified as a by-product. Namely, the metal ores (such as iron, copper, lead, and aluminium) in nature are found in an impure state, often oxidized and mixed with other metal silicates. During ore melting, when ore is exposed to high temperatures, such impurities are separated from the molten metal and can be removed. The collected and removed compounds consist of slag. With once-purified metal, during further processing (casting, alloying), substances are added to melt and enrich it, with reformed slag as a by-product. Therefore, slag mainly consists of ore impurities (mainly silicon and aluminium) combined with calcium and magnesium from various supplements.

Except as a mechanism for removing impurities, during the melting of metals, slag can aid in temperature control during the smelting process and as a reduction method of reoxidation of finished liquid metal before casting. Specifically, the molten metal begins to oxidize and a slag layer forms a protective crust of oxides on the surface, protecting it from further oxidation.

The type of slag formed depends on the type of metal (ferrous or nonferrous) that are processed. Melting processes produce different types of slag, as shown in Figure 1.1. The melting nonferrous metals, iron and silicon, are separated to form a silicon-based slag. The resulting slag contains a high proportion of steel. In contrast, the melting of ferrous metals results in a completely nonmetallic slag because all the steel is used up in the melting process. Such slag mainly contains oxides of calcium, magnesium, and aluminium.

Depending on the cooling and solidification method of the molten masses (those from the processing of ferrous and nonferrous metals), there are a few basic types of slag, shown in Figure 1.2.

Crystalline slag is obtained by casting in a trench and cooling to ambient conditions. Upon mass solidification, cooling can be accelerated by sprays of water, which results in the formation of cracks within the mass and thus facilitates subsequent crushing. This product is mainly crystalline (as indicated by the name), with a cellular or vesicular structure as a result of gas bubbles that formed in the molten mass [18].

Granulated slag is formed by quickly quenching (chilling) molten slag with water or air to produce a glassy state, with little or no crystallisation. After the granulated blast-furnace slag is formed, it must be dewatered, dried, and ground up

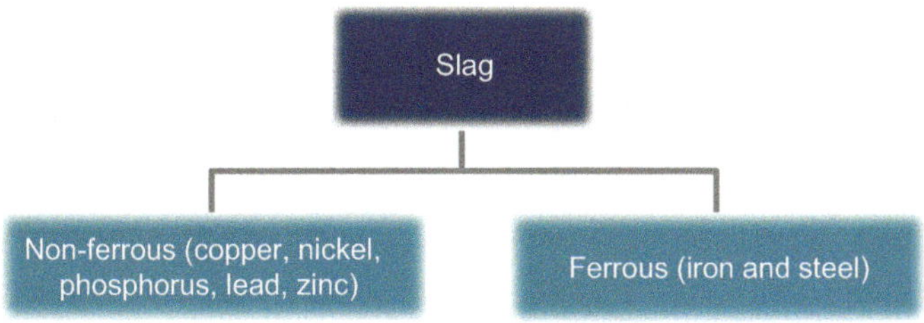

Figure 1.1 Types of slag, according to the melted metal type

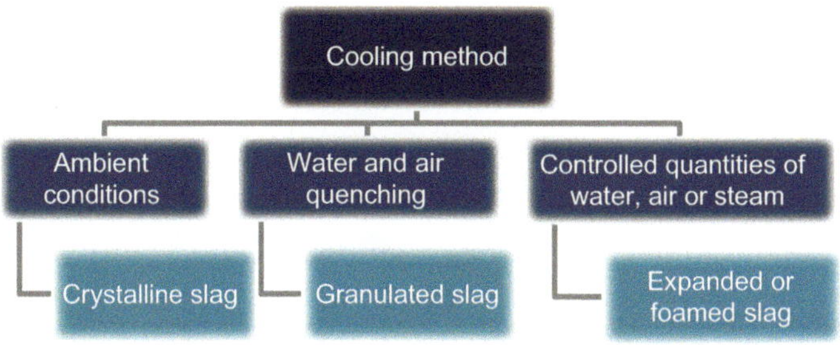

Figure 1.2 Types of slag, according to the cooling method

before it is used as a cementitious material. Magnets are often used before and after grinding to remove residual metallic iron [19]. As a result of this process, sand-size grains and often friable material like clinker are formed. The physical structure and gradation of the resulting slag depends on the chemical composition and temperature of the molten mass during cooling. Sand-size grains resembling dense glass are produced, and they contain oxides that are found in Portland cement, with a significant difference in the proportion of calcium and silicon. Like Portland cement, it has excellent hydraulic properties and, with a suitable activator (such as calcium hydroxide), it will set in a similar manner [18]. The rate of reaction increases with the fineness. Typically, this slag is ground to an air-permeability (Blaine) fineness exceeding that of Portland cement to obtain increased activity at early ages [19].

Expanded or foamed slag results from the treatment of molten slag with controlled quantities of water, air, or foam. Variations in the amount of coolant and the cooling rate will result in variations in the properties of the cooled mass. However, in general, this is a product of a more cellular and vesicular nature than air-cooled slag, and thus is much lighter in weight. Due to the variation in properties, the research literature often cites pelletized slag as a subtype of expanded slag. This slag is generated by a cooling method that involves cooling the molten mass using a limited amount of water, followed by chilling slag droplets thrown through the air by a rapidly revolving finned drum. Depending on the cooling process, the resulting slag particles may be angular and roughly cubical in shape, and thus more appropriate as aggregate, or they may be spherical and smooth, and therefore more suitable for use as a cement additive [18].

Use of a *pelletizer,* also referred to as *air granulation,* involves molten slag passing over a vibrating feed plate, where it is expanded and cooled by water sprays. It then passes onto a rotating, finned drum, which throws it into the air, where it rapidly solidifies into spherical pellets. This slag may also have a high glass content and can be used either as a cementitious material or, for larger particle sizes, as a lightweight aggregate [19].

The most common nonferrous slag are those originated in the processing of copper, nickel, phosphorus, lead, and zinc. The origin of copper and nickel slag can be seen as the result of a multistep process, as shown in Figure 1.3, and lead and zinc slags are formed in a very similar way. After initial processing (grinding), minerals are exposed to temperatures below their melting point. This process, called *roasting,* converts sulphur to sulphur dioxide. Then, reduction of the metal ion via the process of *smelting*

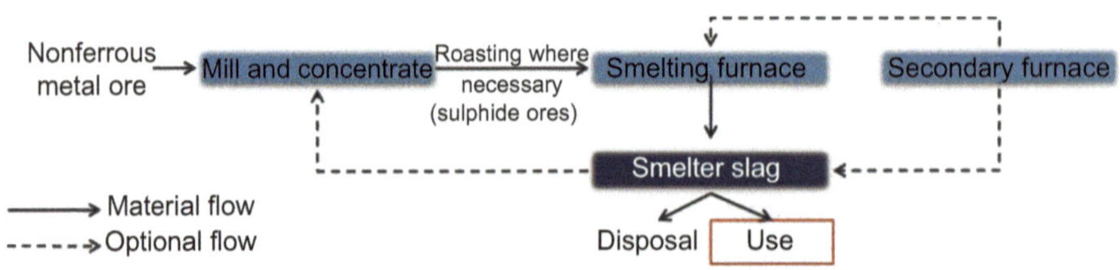

Figure 1.3 Production process of copper, nickel, lead, lead-zinc, and zinc slag

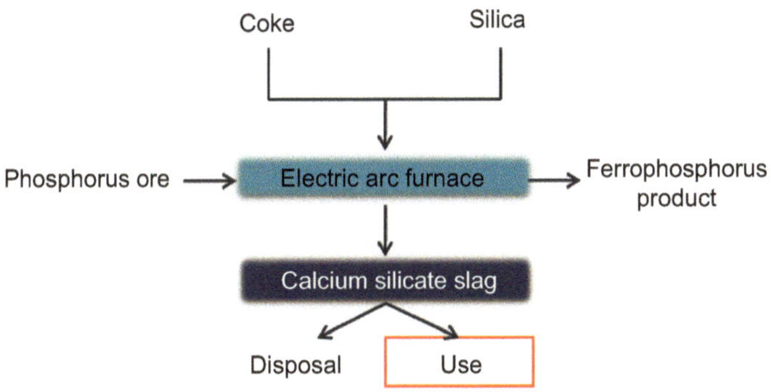

Figure 1.4 Production process of phosphorus slag

is accomplished with the roasted product dissolved in siliceous flux. This melt is then desulfurized with lime flux, iron ore, or basic slag during the process of *conversion*, and then oxygen is lanced to remove other impurities.

Lead, lead-zinc, and zinc slag are formed during pyrometallurgical treatment of the sulphide ores. This process is similar to the production of copper and nickel slag, including roasting, smelting, and conversion.

Phosphorous slag is a by-product of the elemental phosphorus refining process (Figure 1.4). Elemental phosphorus in the EAF is added to flux materials to separate it from the phosphate-bearing rock. The flux additives, whose role during this process is the removal of impurities, are mainly silica and carbon. In addition to silica and carbon, iron can be added in the furnace, which combines with phosphorus to form ferrophosphorus. By the removal of ferrophosphorus (or only phosphorus, if iron is not added), slag is also created.

The amount of nonferrous slag produced in these processes is not as great as ferrous slag. Therefore, researchers have tended to focus their investigations on the larger-volume waste materials. The very few studies that have focused on the basic properties of nonferrous slag and its possible applications in civil engineering are given in Tables 1.2 and 1.3.

Ferrous slag refers to slag generated during the production and casting of iron and steel, as shown in Figure 1.5.

The American Society for Testing and Materials (ASTM) defined blast furnace slag as "non-metallic product consisting essentially of silicates and alumina-silicates of calcium and other bases that is 'developed' in a molten condition simultaneously with iron in a blast furnace" [18]. Such slag consists primarily of impurities of iron ore (mainly silica and alumina) combined with calcium and magnesium oxides from the flux stone. The chemical composition of slag depends on the composition of iron ore, fuel, flux stones, and ratios required for efficient furnace operation.

Steel slag, a by-product of steel production, is generated during the separation of molten steel from impurities in steel production furnaces. Impurities consist of carbon monoxide and silica, manganese, phosphorus, and some iron in a form of liquid oxide. Combined with lime and dolomite-lime, these impurities create steel slag. Since there are three different procedures in steel production, depending on the type of furnace,

Table 1.2 **Physical properties of various nonferrous slag [20]**

Property	Nickel slag	Copper slag	Phosphorus slag	Lead, lead-zinc, and zinc slag
Appearance	Reddish-brown to brown-black	Black	Black to dark gray	Black to red
Texture	Massive, angular, amorphous texture	Glassy; more vesicular when granulated	Air-cooled is flat and elongated; granulated is uniform, angular	Glassy, sharp angular (cubical) particles
Unit weight (kg/m^3)	3,500	2,800–3,800	Air-cooled: 1,360–1,440 Expanded: 880–100	<2,500–3,600
Absorption (%)	0.37	0.13	1.0–1.5	5.0

Table 1.3 **Uses of slags in civil engineering [20–22]**

Type of slag	Use in civil engineering
Copper slag	Aggregate in hot mix asphalt and concrete
Phosphorus slag	Road base and subbase
Zinc, lead-zinc, and lead slag	Aggregate in hot mix asphalt and cement concrete
Tin slag	Aggregate in asphalt layers and unbound granular base layers, mineral wool production

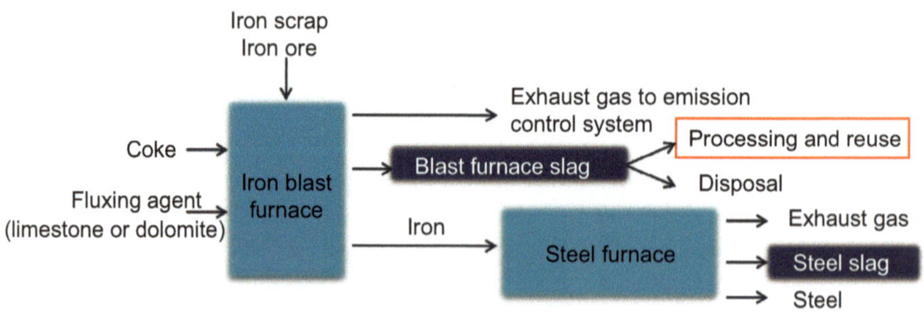

Figure 1.5 Production process of blast furnace and steel slag

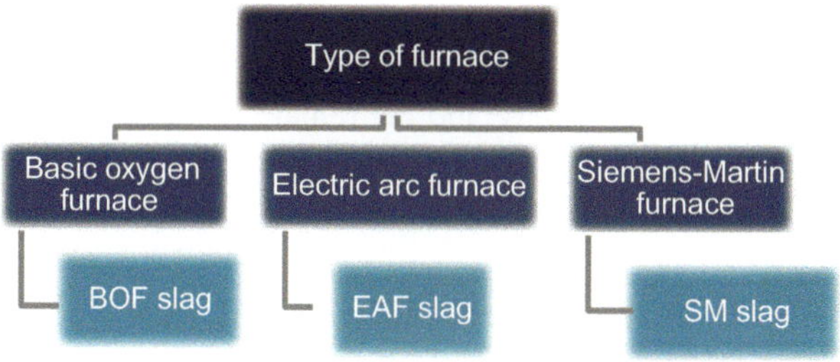

Figure 1.6 Steel slag, produced depending on the steelmaking process

these slags are often referred to by the type of furnace that produced them (Figure 1.6). Steelmaking slag is usually air-cooled.

For steel production, usually one of two processes (furnaces) is used today: BOF or EAF. The BOF process uses 25%–35% old steel (scrap), while the EAF process uses virtually 100% old steel to make new steel. Today in the United States, BOF makes up approximately 40% and EAF makes up about 60% of steelmaking [23].

1.3.3 History of slag utilisation

The British Isles recorded instances of iron processing at the time of the Celts in 700 BC [24], while Aristotle in 350 BC wrote about the use of slag as a drug [25]. The use of slag from iron production was recorded in 1589 by the Germans for making cannon balls [26], and the first use of slag in construction was written about in the context of road base construction during the Roman Empire. The first modern roads with slag were built in England in 1813. By 1880, blocks cast from slag were used for paving streets in Europe and the United States. Since slag was commonly used as ship ballast in that era, it is likely that the *Mayflower,* the ship that transported the Pilgrims from Plymouth, England, to the American colonies in the New World, carried a load of this useful material.

When German businessman Emil Langen discovered the latent hydraulic properties of ground granulated blast furnace slag in 1862, it began to be used widely as a cement additive [27]. About half a century later, in 1909, the first official standard permitting slag to be used in the production of cement was issued in Germany, legally codifying the application of blast furnace slag in cement [28]. Another milestone in the application of slag occurred in 1880, when the application of steel slag as a soil improver was confirmed [27]. Also, historical documents from the 18th century refer to the application of slag in masonry around Europe [26], while other data state that slag cement was used in 1930 during the construction of the Empire State Building [29].

Although slag proved its versatility well before the 20th century, for a long time it was used exclusively as track ballast for railroads in the United States. With the increased production, the need to find new areas of application also grew. One of the first areas of slag application in modern times was in the construction of military roads during World War I. Specifically, political circumstances throughout the world during the 20th century, as well as the rapid development of technology, created a favourable environment for even greater use of this material. Since then, the application of slag has been found useful in many areas, and the remaining chapters of this book present an overview of its possible uses in civil engineering, including building and road construction.

1.4 Concluding remarks

A major promoter of slag as a material of great potential and broad application possibilities in Europe has been made by EUROSLAG, the association of organisations and companies concerned with all aspects of the manufacture and utilisation of slag products. On its website (http://www.euroslag.com/), the organisation regularly publishes statistics on production quantities of slag in Europe, as well as areas in which it is used. Given that iron and steel comprise up to 88% of the metals processed in the world [10], this organisation emphasises ferrous slag (and particularly, its disposal) as major problems. Thus, according to the latest EUROSLAG report [30] for 2012, 23 million tons of blast furnace slag and 21.4 million tons of steel slag were produced; at the same time, the quantity of reused slag was even greater than the amount that was produced. More details on the production and various applications of different types of slag are presented in Chapters 4 and 5 of this book. The same report states that blast furnace slag has found its full application, while some share of steel slag remains unused and ends in dumping sites. For that reason, this type of slag became the subject of extensive research by the authors of this book.

References

[1] I. Dumitru, R. Munn, G. Smorchevsky, Progress toward achieving ecologically sustainable concrete and road pavements in Australia, in: International Conference on the Science and Engineering of Recycling for Environmental Protection, Elsevier Science, Harrogate, UK, 2000.

[2] T. Singh, Green roads: Highways of the future? Accessed August 12, 2015, from http://www.americainfra.com/news/green-roads/, 2012.

[3] World Business Council for Sustainable Development (WBCSD), Report: Recycling concrete. Accessed August 12, 2015, from http://www.wbcsdcement.org/pdf/CSI-RecyclingConcrete-FullReport.pdf.

[4] J.M. Crow, The concrete conundrum, Chemistry World (2008) 62–66. Accessed August 12, 2015, from, http://www.rsc.org/images/Construction_tcm18-114530.pdf.

[5] Department for Environment, Food, and Rural Affairs, Waste legislation and regulations. Accessed August 12, 2015, from https://www.gov.uk/waste-legislation-and-regulations#eu-waste-framework-directive.

[6] European Parliament and Council of European Union, Directive 2008/98/EC-EU Waste Framework Directive, Official J. Eur. Union (2008) L 312.

[7] A. Pappu, M. Saxena, S.R. Asolekar, Solid waste generation in India and their recycling potential in building materials, Bldg. Environ. 42 (2007) 2311–2320.

[8] M. Safiuddin, et al., Utilization of solid wastes in construction materials, Intl. J. Phys. Sci. 5 (2010) 1952–1963.
[9] M. Barbuta, et al., Wastes in building materials industry, Agroecol. (2015). Accessed August 15, 2015, from http://www.intechopen.com/books/agroecology/wastes-in-building-materials-industry.
[10] National Slag Association, Common uses for slag. Accessed March 11, 2015, from http://www.nationalslag.org/common-uses-slag.
[11] A. Hart-Davis, Science: The Definitive Visual Guide [in Croatian], Mozaik knjiga, Zagreb, 2011.
[12] J. A. Ricketts, How a Blast Furnace Works. Accessed August 16, 2015, 2015, from https://www.steel.org/Making%20Steel/How%20Its%20Made/Processes/How%20a%20Blast%20Furnace%20Works.aspx.
[13] T.A. Wertime, The Coming of the Age of Steel, University of Chicago Press, Chicago, 1962.
[14] G. Eggert, What Goes on in the Blast Furnace? Accessed September 29, 2015, from http://www.engr.psu.edu/mtah/essays/blast_furnace.htm.
[15] J. Madias, Treatise on process metallurgy: Industrial processes, Industrial Processes, in: S. Seetharaman, A. McLean, R. Guthrie (Eds.), Electric Furnace Steelmaking, 3, Elsevier, Oxford, UK, 2014.
[16] Y.N. Toulouevski, I.Y. Zinurov, Innovation in Electric Arc Furnaces, Springer, London, 2009.
[17] J.A.T. Jones, Electric arc furnace steelmaking. Accessed August 15, 2015, from https://www.steel.org/making-steel/how-its-made/processes/processes-info/electric-arc-furnace-steelmaking.aspx, 2015.
[18] D.W. Lewis, Properties and uses of iron and steel slags, Presentation at symposium on slag, South Africa, 1982, Accessed October 3, 2015, from www.nationalslag.org/sites/nationalslag/files/documents/nsa_182-6_properties_and_uses_slag.pdf.
[19] S.J. Virgalitte, et al., Ground granulated blast-furnace slag as a cementitious constituent in concrete, Reported by ACI Committee 233 (2000). Accessed January 30, 2015, from http://www.alccofine.com/wp-content/uploads/2012/05/68735435-ACI-233r-95-GGBS-in-Concrete.pdf.
[20] J. De Brito, N. Saikia, Recycled Aggregate in Concrete: Use of Industrial, Construction and Demolition Waste, Springer-Verlag, London, 2013.
[21] M.A. Wahab Yusof, Investigating the Potential for Incorporating Tin Slag in Road Pavements, University of Nottingham, Nottingham, UK, 2005.
[22] A. Van Weers, S. Stokman-Godschalk, Radiation protection, regulatory, and waste disposal aspects of the application of mineral insulation wool with enhanced natural radioactivity. Accessed November 3, 2015, from http://iii.v.8.eu-norm.org/index.pdf.
[23] How Steel Is Made. Accessed October 3, 2015, from http://www.steel.org/making-steel/how-its-made.aspx.
[24] M. Dean, Ferrous slags in UK—The big picture, in: H. Motz, S. Haimi, M. Mäkikyrö (Eds.), 4th European Slag Conference—Proceedings "Slags—Providing Solutions for Global Construction and Other Markets", EUROSLAG Publication, Oulu, Finland, 2005, pp. 19–30.
[25] J. Geiseler, I. Vaittinen, The status of processed slags from iron and steel making, in: J. Geiseler, M. Dean (Eds.), 3rd European Slag Conference—Proceedings "Manufacturing and Processing of Iron and Steel Slags", EUROSLAG Publication, Keyworth, UK, 2002, pp. 37–42.
[26] Slag in history. Accessed October 3, 2015, from http://www.nationalslag.org/slag-history.

[27] H. Motz, Production and use of air-cooled blast furnace and steel slags, in: J. Geiseler, M. Dean (Eds.) 3rd European Slag Conference—Proceedings "Manufacturing and Processing of Iron and Steel Slags", EUROSLAG Publication, Keyworth, UK, 2002, pp. 7–20.
[28] D. Padovani, B. Corcoran, Improved performance of granulated blastfurnace slag (GBS) cements with cement additives. Cemtech Conference 2004. Accessed February 6, 2015, from www.mapei.it/dam/Pdf/ConferencesImproved.pdf.
[29] F. Sajedi, et al., Relationships between compressive strength of cement-slag mortars under air and water curing regimes, Const. Bldg. Mat. 31 (2012) 188–196.
[30] EUROSLAG. Statistics 2012. Accessed February 6, 2015, from http://www.euroslag.com/products/statistics/2012/.

Ferrous slag: characteristics and properties

2

The term *ferrous slag* describes slag that is generated during iron and steel production and casting. Depending on the iron and steel production process, various slag types can be created, generally known as *blast furnace slag* and *steel slag*. Blast furnace slag is made during the melting and reduction of iron ore in a blast furnace, while steel slag is produced during the conversion of hot metal to crude steel or during the melting of scrap in various kinds of furnaces.

Given that the processing of iron and steel makes up to 95% of the total metal processing industry [1], the main waste disposal problem in the industry involves ferrous slag. The characteristics and properties of certain types of ferrous slag, which is important when considering civil engineering applications, are given next.

2.1 Blast furnace slag characteristics

Iron is produced in blast furnaces by the reduction of iron-bearing materials with hot gas. The large, refractory-lined furnace is charged through its top with iron in the form of ore, pellets, sinter, or a combination; flux as limestone, dolomite, and sinter; and coke as fuel. Iron oxides, coke, and fluxes react with the blast air to form molten reduced iron, carbon monoxide, and slag. In the furnace, raw iron is decomposed into molten iron and molten slag when it is melted at a temperature of 1500°C–1600°C. Molten iron is collected at the bottom of the furnace (hearth), while slag floats on the pool of iron.

Around 65% of the world's steel production relies on blast furnaces [2], and blast furnace slag is created in large quantities as a by-product. The American Society for Testing and Materials (ASTM) defines blast furnace slag as a "nonmetallic product, consisting essentially of silicates and aluminosilicates of calcium and other elements that are formed in solution along with iron in the blast furnace" [3]. Such slag consists primarily of impurities of iron ores (mainly silicon and aluminium) combined with calcium and magnesium oxides of flux.

2.1.1 Physical-mechanical and chemical properties of blast furnace slag

The technical properties of blast furnace slag used in civil engineering depend on its chemical and physical properties. The chemical composition of slag depends on the composition of iron ore, fuel, and flux, as well as the ratios required for efficient

Characteristics and Uses of Steel Slag in Building Construction. http://dx.doi.org/10.1016/B978-0-08-100368-8.00002-6

operation of the furnace. Furnaces must be charged with uniform feedstock to produce iron that is uniform in quality. This process also results in uniformity (within very narrow limits) for slag composition and chemical composition. Therefore, it can be said that 95% of blast furnace slag is composed of four main oxides: calcium, magnesium, silicon, and aluminium [3,4].

When melted at 1500°C–1600°C, blast furnace slag has about 30%–40% SiO_2 and about 40% CaO, which is close in composition to Portland cement [5], the most widely used binder in civil engineering. Consequently, the reactions of activated blast furnace slag are analogous to those of Portland cement. Besides the oxides, blast furnace slag contains sulphur (originating from the coke), iron, manganese, and other trace elements. Table 2.1 presents the typical chemical composition of blast furnace slag.

Depending on the cooling method used for melted slag, different mineralogical compositions are created. Figure 2.1 presents the typical minerals of air-cooled blast furnace slag.

Table 2.1 Major chemical constituents in blast furnace slag [3,6]

Constituent	Wt %
Lime (CaO)	32–45
Magnesia (MgO)	5–15
Silica (SiO_2)	32–42
Alumina (Al_2O_3)	7–16
Sulphur (S)	1–2
Iron oxide (Fe_2O_3)	0.1–1.5
Manganese oxide (MnO)	0.2–1
Iron (II) oxide (FeO)	≤ 1
Chromium (III) oxide (Cr_2O_3)	≤ 1

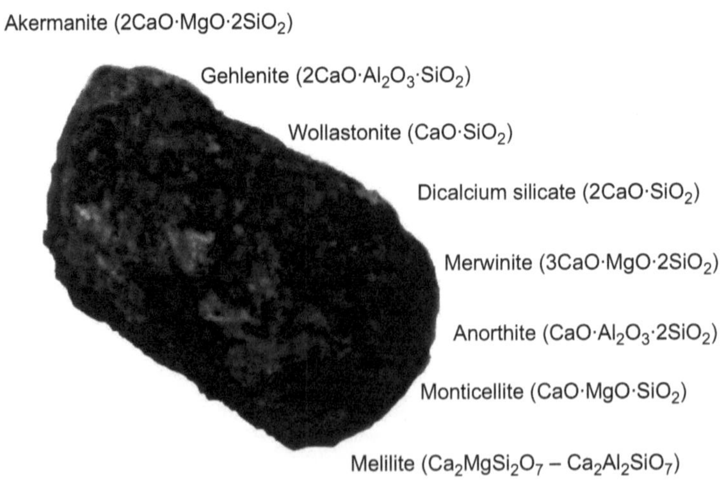

Figure 2.1 Minerals of air-cooled blast furnace slag [3]

Slag that is cooled rapidly after emerging from the furnace tends to form a glassy, noncrystalline structure, while slower cooling leads to crystallisation of a number of minerals, depending upon the relative proportion of major oxides in the slag [3]. Generally, melilite ($Ca_2MgSi_2O_7$–$Ca_2Al_2SiO_7$) and merwinite ($3CaO{\cdot}MgO{\cdot}2SiO_2$) are the most common minerals presented in blast furnace slag, and most slags contain no more than four minerals depending on their chemical composition [3].

In addition to the influence of raw materials on the physical-mechanical and chemical properties of slag, maintenance of blast furnace slag is a significant issue. To protect the inside of the furnace against premature erosion and wear (especially below the taphole or at the corners of the hearth), titanium dioxide (TiO_2) is used [2]. The addition of TiO_2 to the burden has negative physical effects, such as higher porosity, higher residual moisture, darker colour, and lower reactivity, with a lower compressive strength of slag cement [7]. When the content of TiO_2 in slag exceeds 25%, it becomes ropy and foamy [8]. The presence of titanium oxide, as with titanium carbide and titanium nitride, induces changes in slag surface characteristics (surface tension and viscosity), which are favourable for stabilising foam in expanded or foamed blast furnace slag. However, large quantities of gas are produced, which is unfavourable, so titanium oxide can be reduced by operating with a low temperature and high oxygen potential in the blast furnace or with the generation of carbon monoxide.

Like any other slag, blast furnace slag may be crystalline, granular, or expanded, depending on how it was cooled. Over 90% of blast furnace slag is air cooled [3]. This method is very simple, but also time consuming; liquid slag is poured into beds and slowly cooled under ambient conditions. Under these conditions, a crystalline structure forms, which is then crushed and screened. Crushed air-cooled slag is sharp and cubical and has a surface texture ranging from rough and porous to glassy. Depending on the process of steel production, there are variations in the physical properties of slag. Generally, the density of air-cooled slag is greater than the density of the granular or foamy slag, but more specifically, the properties will depend on the metal being processed and details of the cooling process. This slag also has high hardness (5–6 on the Mohs scale) and lower thermal conductivity than ordinary aggregate. Due to its physical properties, this slag is commonly used as an aggregate in asphalt in concrete mixtures. As an aggregate, it is characterised by high resistance to freeze/thaw cycles and high resistance to abrasion. Added to concrete as a cement additive, this slag will increase the workability time of fresh concrete mixture. In addition to being used as an aggregate in concrete and asphalt or as a cement additive, this slag is used for constructing road substructure, as infill material, as railbed material, and as a filter in sewerage systems and their covering material and drainage layers [9]. The high stability of air-cooled slag aggregate and its ability to lock up in granular base applications make it a valuable material for construction over soft ground.

As already stated, the air-cooling method is time consuming, so accelerated cooling is undertaken by introducing controlled quantities of water, air, or steam. This cooling method results in expanded or foamed blast furnace slag. It is a sharp, cubical, and coarse-textured material (which is coarser than air-cooled slag). Its density is about 70% of the density of air-cooled slag [9]. Because it has lower density, it also has greater porosity. The main characteristic of this slag is its foamy structure, which is created by two main factors [8]: (i) the speed of gas production in slag during its cooling, and (ii) whether slag performance is available for gas to discharge. Also, slag surface characteristics, especially surface tension and surface viscosity, affect the process of foaming. Foam generation is directly affected by slag surface tension, while foam stabilisation is affected by slag surface viscosity.

Expanded blast furnace slag can be used as a lightweight aggregate in concrete, concrete blocks, and precast units. Nearly 90% of the resulting expanded slag is used in concrete as lightweight aggregate; compared with other aggregates, it provides the concrete with better thermal insulation and fire resistance [3]. It is also used as a cement additive, material for drainage, and as lightweight filling [9].

If a rolling or spinning drum is introduced into the blast furnace cooling process, with water and air as cooling agents, pelletised blast furnace slag is produced. Pellets can be more vitrified when rapid quenching is used, or more crystalline with slower quenching [10]. Pelletised blast furnace slag has a smooth texture, sealed surface, and rounded shape, unlike air-cooled or expanded blast furnace slag [9,10]. These physical properties result in the slag having lower porosity and water absorption than air-cooled and expanded slag, as well as improved workability and a lower cement requirement when used as an aggregate in concrete. It also has a greater bulk density than expanded slag. Its vitreous state results in relatively lower thermal conductivity than that of ordinary lightweight aggregate.

The production process defines the characteristics and potential uses of pelletised blast furnace slag. Crystalline pellets are used as a lightweight aggregate, in lightweight concrete blocks or structural concrete, or for cement production. Usage for basement insulation purposes has also been described [10].

When uncontrolled quantities of cooling agent are used, granulated blast furnace slag is produced. This is a glassy, granular material that is formed when molten blast furnace slag is rapidly chilled, such as by immersion in water as the slag exits the blast furnace. Modern technologies have provided a more efficient technique, which uses high-pressure water jets that impinge on the stream of molten slag at a water-slag ratio of about 10:1 by mass [11]. This rapid cooling method minimises mineral crystallisation, resulting in fine, sand-sized particles composed mainly of glass.

After cooling, this slag is crushed, graded, or ground depending on the desired application. The grinding properties of the slag depend on its origin and chemical composition, which yield different levels of grindability and wear rates for the equipment used [12]. Namely, during the production of blast furnace slag, some metallic iron is entrapped in it (so-called pig iron), resulting in high grinding-element wear. Because of the abrasive properties, grinding parts of the rollers used for blast furnace slag grinding have to be hardfaced.

Dried and ground glass in this type of slag are latently hydraulic, which means that they do not need lime to hydrate—just an alkaline environment. While the slag contains the same major oxides as Portland cement, glass in granulated blast furnace slag has different proportions of lime and silica. That is why hydration properties of granulated blast furnace slag depend on a glass content to a large extent. When mixed with water, this slag reacts by forming cementing hydration products. The strength of these cementing reactions will depend on the chemical composition of slag, glass

Table 2.2 Typical physical properties of blast furnace slag [9]

Property	Air-cooled slag	Expanded slag	Pelletised slag
Specific gravity	2.0–2.5	–	–
Compacted unit weight (kg/m^3)	1120–1360	800–1040	840
Absorption (%)	1–6	–	–

Table 2.3 Comparison of mechanical properties of air-cooled blast furnace slag coarse aggregate and natural aggregate [9,14]

Property	Slag aggregate	Natural aggregate
Los Angeles abrasion test	35%–45%	15%–45%
California bearing ratio	>100	80–100
Mohs hardness	5–6	3–8
Sodium sulphate soundness loss	12%	–
Angle of friction	40–45 degrees	30–45 degrees

content (as stated previously), and the fineness of the grinding. For example, Al_2O_3 (up to 13% in slag chemical composition) and MgO (up to 11%) increase the strength [13]. However, the presence of manganese (Mn) reduces slag hydraulic activity but stabilises the glassy phase. Namely, a high level of MnO is known to inhibit early-age hydration of cement but has no negative influence on long-term properties [13].

Chemical reactions of granulated blast furnace slag with water are slow and can be accelerated by the presence of calcium hydroxide, alkali, and cement. Because of its cementing properties, granulated blast furnace slag is often used as a cement additive, to be added directly to Portland cement or to hydrating limestone to produce a blended cement. It can also be added to Portland cement concrete as a mineral supplement [9]. The typical physical properties of blast furnace slag are given in Table 2.2, and the mechanical properties are described in Table 2.3.

2.1.2 Potential corrosive effect

Another important material property that affects the use of slag in civil engineering is pH value, which dictates the potential corrosive effect if slag is used in concrete. Blast furnace slag is mildly basic (alkaline), with a pH in solution of 8–10 [9]. Although it contains 1%–2% of elemental sulphur, the leachate is slightly alkaline and does not present a risk of corrosion for steel embedded in concrete that contains slag cement or slag aggregate. However, where blast furnace slag is in contact with stagnant water, discoloration may sometimes occur (typically a yellow/green colour) and an intense sulphur odour may be produced. Under these conditions, the stagnant water has a high concentration of calcium and sulphur and a pH value of about 12.5 [15]. When exposed to oxygen, sulphides within this yellow leachate react to precipitate elemental sulphur, producing a calcium thiosulphate (transparent solution). Ageing of slag may delay the occurrence of yellow leachate, but there is no guarantee that it will not occur if stagnant water is left in contact with blast furnace slag for a long period of time [4,9].

2.1.3 Decomposition of dicalcium silicate and iron

Besides the problems related to the presence of sulphur, another possible problem in using blast furnace slag in civil engineering is its disintegration. Slow-cooled (air-cooled) blast furnace slag with a high amount of lime may form dicalcium silicates, which are prone to increasing their volume during cooling at an ambient temperature. This volume increase results in the decomposition of slag into powder, making its handling and storage problematic and giving it virtually no economic value. Decomposition occurs prior to slag being used in construction, however, and it will not present a problem for the end user, but rather for the slag manufacturer [3,16]. This problem can be avoided by changing the chemical composition or by using a liquid slag cooling method. Thus, it is recommended that blast furnace slag be exposed to rapid cooling in order to prevent crystallisation of dicalcium silica [3]. In addition to dicalcium silica, slag disintegration can be caused by a higher proportion of iron oxide, which, paired with certain amounts of other constituents, can form components that will react with water and cause blast furnace slag disintegration.

2.2 Steel slag characteristics

Steel slag is a by-product of molten iron processing, and different types of steel slag are formed depending on a specific type or grade of steel and the furnace used during steel production. Unlike blast furnace slag, which is classified into different types depending on the cooling method, practically all steel slag is cooled naturally, by air under ambient conditions. Steel slag is classified depending on the furnace used for steel production. The basic oxygen furnace (BOF) produces BOF steel slag, the electric arc furnace (EAF) produces EAF steel slag, and the ladle (LD) furnace produces LD steel slag.

2.2.1 Physical-mechanical and chemical properties

The composition of steel slag varies depending on the furnace type, composition of charges, and grades of produced steel. The typical chemical composition of steel slag is presented in Table 2.4.

Since BOF and EAF steel slags are basic results or by-products of the steelmaking process, they have a similar chemical and mineralogical composition, which includes two major compounds: calcium oxide and iron oxide. With regard to the chemical composition of LD slag, alloys added during production of different grades of steel have major influences, so it differs from those of BOF and EAF slag. The typical chemical composition of EAF, BOF, and LD slags is given in Table 2.5.

The typical mineral composition of EAF, BOF, and LD slags is given in Table 2.6.

Table 2.4 **General chemical composition of steel slag [6,17]**

Constituent	Weight (%)
Lime (CaO)	35–45
Magnesia (MgO)	2–9
Silica (SiO_2)	11–17
Alumina (Al_2O_3)	1–6
Sulphur (S)	≤ 0.2
Iron oxide (Fe_2O_3)	0.9–38
Manganese oxide (MnO)	2–6
Iron (II) oxide (FeO)	16–26
Chromium (III) oxide (Cr_2O_3)	0.5–2
P_2O_5	Approx. 0
TiO_2	0.4–3.0

Table 2.5 **Typical chemical composition of EAF, BOF and LD slag (Wt. %) [17,18]**

Constituent	EAF	BOF	LD
Lime (CaO)	23.9–60	30–60	30–60
Magnesia (MgO)	2.9–15	0.8–13	1–12.6
Silica (SiO_2)	9–20	7.8–20	2–35
Alumina (Al_2O_3)	2–9	0.98–5	5–35
Sulphur (S)	< 0.25	Approx. 0.25	Approx. 0.25
SO_3	0.1–0.6	0.1–0.8	0–0.3
Iron oxide (Fe_2O_3)	20.3–32.56	38.06	0.9–3.3
Manganese oxide (MnO)	2.5–8	0.3–4.3	0.2–1.4
Iron (II) oxide (FeO)	5.6–34.4	7–26.3	0–15
Chromium (III) oxide (Cr_2O_3)	< 2	< 2	< 2
P_2O_5	0–1.2	0.2–4	0–0.4
TiO_2	0.56–0.8	0.5–2	0.3–0.9

During the conversion of iron into steel in a BOF, a certain share of iron will not be recovered in the produced steel, and it will be incorporated within the BOF chemical composition. For that reason, the content of iron oxide (FeO or Fe_2O_3) can reach 38%. Other chemical compounds in BOF include silica (SiO_2), ranging from 7%–18%; Al_2O_3, from 0.5%–4%; MgO, from 0.4%–14%; CaO, more than 35%; and free lime, up to 12% [17].

The EAF steelmaking process uses a high amount of recycled steel, so its influence on the chemical composition of EAF slag is significant and can vary much more than the chemical composition of BOF slag. The main compounds in EAF slag are FeO, ranging from 10%–40%; CaO, from 22%–60%; SiO_2, from 6%–34%; Al_2O_3, from

Table 2.6 **Typical mineral composition of EAF, BOF, and LD slag (wt.%) [19,20]**

Mineral	EAF	BOF	LD
Tricalcium silicate—$3CaO{\cdot}SiO_2$	0	0	—
Dicalcium silicate—$2CaO{\cdot}SiO_2$	0	35–70	Present
Dicalcium ferrite—$2CaO{\cdot}Fe_2O_3$	0	0–30	Present
Spinel $MeO{\cdot}Me2O_3$ (general formula)	5	0–20	—
Wuestite—FeO·	60	10–20	Present
Free lime—CaO	0	0–10	-
Bredigite—$1.7CaO{\cdot}0.3MgO{\cdot}SiO_2$	10	0	—
Gehlenite—$2CaO{\cdot}Al_2O_3{\cdot}SiO_2$	0	25	-

3%–14%; MgO, from 3%–13%; and some other minor elements (namely, oxidised impurities like MgO, MnO, and SO_3) [17].

Due to the specific process in which LD slag is produced, its chemical composition is highly variable, depending on the alloys fed into the LD furnace and the grade of steel that is produced. However, common chemical elements present in BOF and EAF slags are also present in LD slag: Al_2O_3 and CaO content is usually higher than that in EAF and BOF slag, while the FeO content is up to 10% [17].

The formation of all kinds of minerals in slag is influenced by the cooling rate and slag chemical composition. For blast furnace slag, rapid cooling results in a noncrystalline structure with a high amount of glass, which is responsible for its hydraulic activity. In contrast, low silica content in steel slag results in a crystalline phase, even with rapid cooling methods. High iron oxide content in steel slag (mainly in BOF and EAF slag) makes wustite $(Fe_{1\ -x-\ y},Mg_x,Mn_y)O_z$ one of its main minerals. Other main mineral constituents of steel slag are dicalcium silicate ($2CaO{\cdot}SiO_2$), dicalcium ferrite ($Ca_2Fe_2O_5$), merwinite ($3CaO{\cdot}MgO{\cdot}2SiO_2$), and olivine ($2MgO{\cdot}2FeO{\cdot}SiO_2$) [6,17]. In LD slag, polymorphas of C_2S are frequently observed and are responsible for its high decomposition and expansive nature [17].

Physical-mechanical properties of steel slag are represented by its high angular shape and rough surface texture. Steel slag has high bulk specific gravity and moderate water absorption. Low impact value, high compressive strength, good polishing, and freeze/thaw resistance [6] are also very favourable steel slag characteristics that permit its main application as a substitute for natural rock or gravel materials. The absence of a glassy phase in steel slag also presents an opportunity for its main application, as an aggregate in a broad range of civil engineering fields.

If steel slag is used as an aggregate, special attention should be paid to its mineralogical form due to its potentially expansive nature. Before utilisation as an aggregate, it is recommended that steel slag be washed out. Additionally, the use of slag containing more than 3% impurities and containing components of soft lime is not recommended [21].

Application of steel slag in Portland cement concrete is generally not common because of its alleged corrosive effects on metal parts embedded within reinforced concrete elements. Specifically, a small quantity of sulphur contained in slag is generally

Table 2.7 Typical physical properties of steel slag [21]

Property	Steel slag
Specific gravity	3.2–3.6
Compacted unit weight	1600–1920 kg/m^3
Absorption	Up to 3%

Table 2.8 Comparison of mechanical properties of steel slag and natural aggregate [18, 21]

Property	Slag aggregate	Natural aggregate
Los Angeles abrasion test	20%–25%	20%–40%
California bearing ratio	Up to 300%	100%
Mohs hardness	6–7	3–4
Sodium sulphate soundness loss	< 12%	—
Angle of friction	40–50 degrees	—

considered to be of concern [4]. This belief is based on the fact that some types of fly ash promote the corrosion of metal parts in construction, which is attributed to the presence of sulphur in their composition. However, these types of fly ash were mainly composed of silica, alumina, and iron oxides, which all form acids in the presence of moisture. Calcium and magnesium oxide, which form an alkaline environment, are found substantially less in fly ash. Studies conducted in the past few decades have elucidated the conditions of corrosive effects of sulphur—sulphur is active only in an acidic environment, which encourages corrosion [4]. However, in the composition of steel slag, calcium and magnesium oxide are represented the most, which would create an alkaline environment in concrete and prevent the corrosion of metal embedded in the concrete. The reason why slag is still considered corrosive is mainly economic—it is easier to declare slag as unfit than to investigate its potentially corrosive effects [4]. The typical physical properties of steel slag are given in Table 2.7, and the mechanical properties in Table 2.8.

2.2.2 *Volume (In)stability*

Potential volume changes in steel slag are attributed to the content of free calcium and magnesium oxide (free CaO and MgO) in its chemical composition [3,22]. Free CaO and MgO hydrate under the influence of moisture, which leads to large volume changes. In doing so, free CaO hydrates rapidly, causing large volume changes in the first few weeks. MgO hydrates slowly, contributing to long-term expansion that can extend to several years, even when using aged slag [22]. In Figure 2.2, the

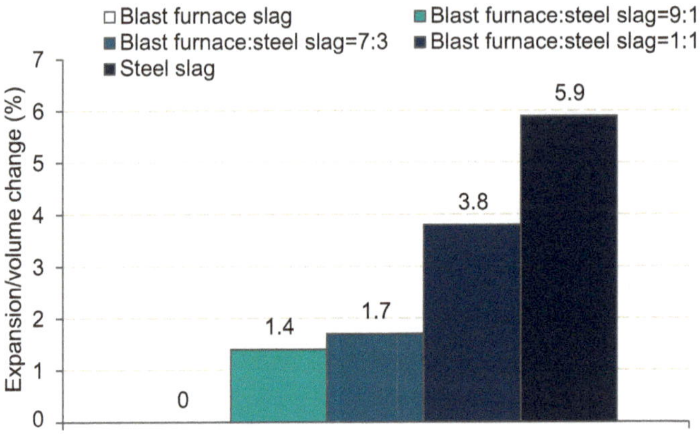

Figure 2.2 Volume change as a function of the blast furnace–to–steel slag ratio [23]

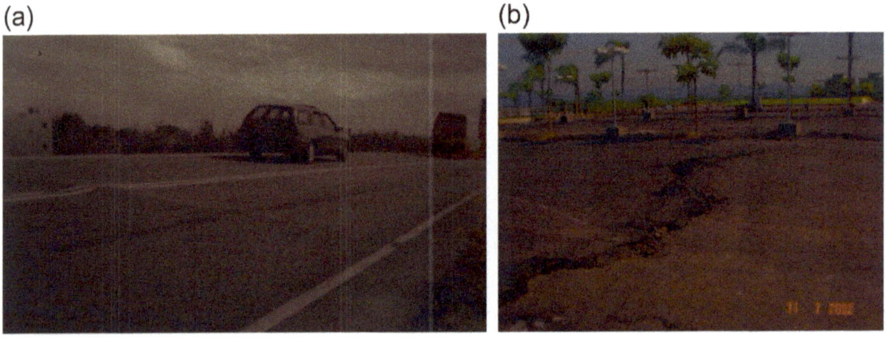

Figure 2.3 Consequences of volume-unstable slag inbuilt in an asphalt mixture; (a) road bump [24]; (b) asphalt-concrete collapse [25]

expansive nature of blast furnace and steel slag is presented with the variation in the blast furnace–to–steel slag ratio.

This potentially expansive nature (i.e., volume changes up to 10% that were associated with the hydration of free CaO and MgO) can cause difficulties with products that contain such slag, which is one of the main reasons that it is considered inappropriate for use in Portland cement concrete or in concrete slab foundations [21].

The consequences of using volume-unstable slag inbuilt in construction are shown in Figures 2.3 and 2.4. Figure 2.3 shows the problem with asphalt concrete in the initial stage (Figure 2.3a) and final stage (Figure 2.3b) of destruction caused by volume instability of slag. Figure 2.4 shows a school building in which the soil moisture induced the expansion of slag used as filling; slag removal was necessary to avoid a collapse of the entire structure.

To prevent these and similar problems (before use of slag in concrete, for example), verification of its volume instability is required in accordance with current legislation.

Slag ageing when exposed to atmospheric conditions for a long period of time (i.e., weathering) is the simplest method of controlling volume instability, but it is not necessarily efficient. In addition to weathering often being a time-consuming process,

Figure 2.4 Removal of slag applied as fill [25]

(a) (b)

Figure 2.5 Accelerated hydration of instable components in slag; (a) slag splashing; (b) slag mixing [25]

it does not even ensure that expansion will be effectively prevented. For instance, rain reacts with free CaO and SiO_2 to form a thin "protective" layer at the exposed surface. This layer prevents ageing of the slag inside the landfill. Therefore, occasional mixing of disposed material is necessary [25] (Figure 2.5).

The time required for slag ageing is individual and varies depending on its type and application; for example, the amount of free CaO and MgO will affect the time needed for ageing. Therefore, sometimes it may take only a few months of weathering or periodical spraying with water. According to Belgian and Dutch regulations, a 1-year ageing period is necessary for use in unbounded base layers of pavement, while an ageing period of up to 18 months for use as an aggregate has also been reported [21].

Although the use of slag in concrete is rejected in most cases in which slag volume instability is proved, that is not necessary. In fact, there are some nonstandard but effective methods for accelerated slag ageing.

Figure 2.6 Accelerated slag ageing by steam exposure under a tarpaulin [25]

Ageing by steam was developed in Japan. One procedure consists of covering slag with a tarpaulin or sheet (Figure 2.6), injecting steam for a period of 48 h, uncovering the slag, and allowing natural cooling [25]. The entire procedure lasts 6 days. Another method of accelerated ageing involves exposing the slag to steam at a pressure of 0.5 MPa for a period of 3 h [25]. Both procedures produce similar results—that is, a greatly reduced number of slag volume changes. In addition to suppressing volume instability, corrections in the process of slag formation can be avoided, even the formation of instable components in a melted mass. According to research [25], adding silica sand in the liquid slag and blowing oxygen prevents the formation of free CaO and MgO.

The issue of forecasting the steel slag expansion ratio was reported by Wang [26], who gave the following expression for predicting the total volume changes:

$$E_s = 0.38 \times \gamma_s \times F \tag{1}$$

where E_s stands for steel slag volume expansion (%), γ_s is compacted density (g/cm^3), and F is free lime content of the steel slag (%).

To confirm the accuracy of Eq. (1), the author examined the volume expansion of BOF slag from three different suppliers, in conditions with and without the additional load (surcharge) that is weighing layers above the layer containing slag. The results of this experiment are shown in Figure 2.7. In this research, it was concluded that Eq. (1) for the prediction of volume change fit the test results well. It was also concluded that the layer under an additional load of 25 g/cm^2 (which is approximately equal to a load from a 10-cm-thick asphalt layer above the layer with slag) reduced the expansion by 7%–8%. This means that if the expansion assumed by Eq. (1) is less than 7%–8%, no volume change will accrue due to the influence of the load (expansion will be annulled through the voids in the material).

Volume changes limit the use of steel slag in rigid pavement structures. But they can be controlled (in asphalt mixtures) or can even improve the properties of embedded material if used in shoulders or unpaved parking areas.

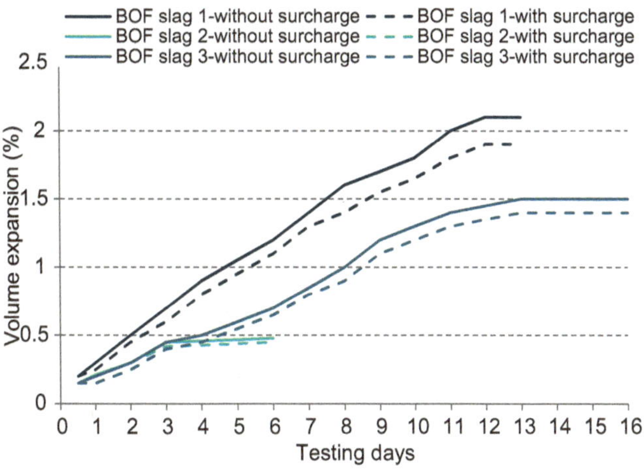

Figure 2.7 Volume expansion results for three BOF slags with and without an additional load (i.e., surcharge) [26]

In addition, due to the exposure of slag aggregates to water and weathering, calcium carbonate is formed, which produces a white sediment. It is precipitated in the form of a white powder and can block the drainage system and cause water retention. These blockages are especially dangerous if freezing occurs, which leads to major pavement damages. The possible formation of such sediment is attributed to slag containing more than 1% CaO [21], and this phenomenon, as opposed to the expansion, cannot be prevented by ageing.

Another chemical element present primarily in LD slag is responsible for volume changes up to 10%. The dicalcium silicate ($2CaO{\cdot}SiO_2$) phase is stable at temperatures higher than 630°C, while at temperatures below 500°C, it starts its phase transformation, causing volume changes [17]. An additional negative consequence of $2CaO{\cdot}SiO_2$ instability is the breakage of crystals and dust emergence during slow cooling of LD slag.

2.2.3 Decomposition of dicalcium silicate and iron

Compared to blast-furnace slag, steelmaking slag usually contains a much higher amount of lime, which can cause formation of dicalcium silicate, $2CaO\text{-}SiO_2$ (sometimes formulated as $2CaO{\cdot}SiO_2$), which can cause disintegration upon cooling due to a volume increase when changing from one crystalline form to another (from the β form to the γ form) [3]. This transition from β to γ form is accompanied by an increase in volume of around 12%, which results in the decomposition of slag into powder [27]. According to Mombelli et al. [27], formation of only 4 wt.% of $\gamma\text{-}Ca_2SiO_4$ is enough to cause slag disintegration. The decomposition of dicalcium silicate is shown in Figure 2.8.

Based on actual experience, if there is a danger of dicalcium silicate decomposition, it will occur prior to material being placed in construction. Therefore, it does not pose a problem for the end user [3]. However, altering the chemical composition of the slag and rapid cooling of the molten mass while preventing the crystallisation of dicalcium silicate can completely prevent this problem.

Figure 2.8 Decomposition of dicalcium silicate [28]

The problem of iron decomposition is considered to be rare and characteristic for slag with a high content of iron oxide [3]. Such slag can, with a certain amount of other constituents, form compounds that will easily react with water and thus lead to the disintegration of material. However, decomposition of the β form of dicalcium silicate (also called *larnite*) in steel slag can be avoided by the addition of melted quartz in the slag flow [27]. Namely, quartz addition has a twofold effect: it reacts with calcium aluminates to form gehlenite, which inhibits the formation of larnite, and it simultaneously prevents its disintegration, thus avoiding the so-called dusting effect.

2.3 Stainless steel slag

According to data from the International Stainless Steel Forum [29], the worldwide demand for stainless steel is increasing, and production reached 41.7 million tonnes in 2014 alone. Along with increasing stainless steel production, the amount of slag formed as a by-product is growing.

Stainless steel slag includes EAF, argon oxygen decarburisation (AOF), and ladle metallurgy (LM) slag. The last two types of slag are generated during the basic refining process of making stainless steel. Although the chemical composition of these types of slag is highly variable, both AOD and LM slag is rich in Ca, Si, and Mg. The main minerals typically found in these various slags are dicalcium silicate ($2CaO{\cdot}SiO_2$), merwinite [$3CaO{\cdot}MgO{\cdot}2SiO_2$], bredigite [$1.7CaO{\cdot}0.3MgO{\cdot}SiO_2$], and periclase (MgO) [30]. Dicalcium silicate undergoes a series of polymorphic transformations upon cooling, one of which is transformation from its β to its γ form, which leads to the disintegration of slag. According to [31, 32], the potential routes to avoid the formation of γ-dicalcium silicate are chemical stabilisation by the addition of boron (B_2O_3) or nonboron compounds (MgO, Na_2O, K_2O, BaO, MnO_2, and Cr_2O_3), changes in slag chemistry, and fast cooling.

References

[1] Iron and Steel Statistics and Information. Accessed January 25, 2015, from http://minerals.usgs.gov/minerals/pubs/commodity/iron_&_steel/.

[2] J. Mochón, et al., Protection mechanisms for blast furnace crucible using titanium oxides, Metall. Mat. Eng. 18 (3) (2012) 195–201.

[3] D.W. Lewis, Properties and uses of iron and steel slags. Presentation at symposium on slag, South Africa. 1982. Accessed January 12, 2015, from www.nationalslag.org/sites/nationalslag/files/documents/nsa_182–;6_properties_and_uses_slag.pdf.

[4] Slag and its relation to the corrosion characteristics of ferrous metals. National Slag Association Report NSA 172-13. Accessed January 14, 2015, from http://www.nationalslag.org/sites/nationalslag/files/documents/nsa_172-13_slag_and_corrosion.pdf.

[5] R.D. Hooton, Canadian use of ground granulated blast-furnace slag as a supplementary cementing material for enhanced performance of concrete, Canad. J. Civil Eng. 27 (2000) 754–760.

[6] EUROSLAG. Properties. Accessed January 15, 2015, from http://www.euroslag.com/products/properties/.

[7] A. Ehrenberg, Granulated blastfurnace slag - State of the art and potentials for the future, in: H. Motz, P. Verhaag (Eds.), 6th European Slag Conference:Ferrous Slag—Resource Development for an Environmentally Sustainble World, EUROSLAG Publication, Madrid, Spain, 2010, pp. 279–299.

[8] C. Bai, G. Qiu, H. Pei, Research on the foaming property of blast furnace slag bearing TiO_2, in: C. Pistorius (Ed.), VII International Conference on Molten Slags, Fluxes, and Salts: The South African Institute of Mining and Metallurgy, Johannseburg, South Africa, 2004, pp. 483–486.

[9] User guidelines for byproducts and secondary use materials in pavement construction. Blast furnace slag. Accessed January 18, 2015, from http://www.fhwa.dot.gov/publications/research/infrastructure/pavements/97148/bfs1.cfm.

[10] J. J. Emery, Pelletized lightweight slag aggregate. 1980. Accessed January 19, 2015, from http://www.nationalslag.org/sites/nationalslag/files/documents/nsa_180-1_pelletized_ltwt_slag_as_aggregate.pdf.

[11] S.J. Virgalitte, et al., Ground Granulated Blast-Furnace Slag as a Cementitious Constituent in Concrete, Reported by ACI Committee 233 (2000). Accessed January 30, 2015, from http://www.alccofine.com/wp-content/uploads/2012/05/68735435-ACI-233r-95-GGBS-in-Concrete.pdf.

[12] C. Hackländer-Woywadt, Grinding of granulated blastfurnace slag in Loesche vertical roller mills, in: H. Motz, P. Verhaag (Eds.), 5th European Slag Conference: Slag Products—Providing Sustainable Solutions for the Built Environment, EUROSLAG Publication, Luxembourg, Luxembourg, 2007, pp. 181–192.

[13] J. Péra, J. Ambroise, M. Chabannet, Properties of blast-furnace slags containing high amounts of manganese, Cement Concr. Res. 29 (1999) 171–177.

[14] Use of Air-Cooled Blast Furnace Slag as Coarse Aggregate in Concrete Pavements. Accessed January 25, 2015, from http://www.fhwa.dot.gov/pavement/concrete/pubs/hif12031.pdf.

[15] Blast Furnace Slag. Accessed January 25, 2015, from http://ispatguru.com/blast-furnace-slag/.

[16] A.M. Dunster, Blast furnace slag and steel slag as aggregates: A review of their uses and applications in UK construction, in: J. Geiseler, M. Dean (Eds.), 3rd European Slag Conference: Manufacturing and Processing of Iron and Steel Slags, EUROSLAG Publication, Keyworth, UK, 2002, pp. 21–29.

[17] I.Z. Yildirim, M. Prezzi, Chemical, Mineralogical, and Morphological Properties of Steel Slag. Advances in Civil Engineering, 2011, Accessed January 30, 2015, from http://www.google.hr/url?sa=t&rct=j&q=&esrc=s&source=web&cd=2&ved=0CCwQFjAB&url=http%3A%2F%2Fdownloads.hindawi.com%2Fjournals%2Face%2F2011%2F463638.pdf&ei=JlLLVI35Lsa4UZTegpAP&usg=AFQjCNF_YDHaH-54Jp1zjMZ_cwRhj25_Gqg&bvm=bv.84607526,d.d2s.

[18] P. Miklos, The utilization of electric arc furnace slags in Denmark, in: J. Geiseler, M. Dean (Eds.), 2nd European Slag Conference—Engineering of slags: A scientific and technological chalenge, EUROSLAG Publication, Düsseldorf, Germany, 2000, pp. 101–110.

[19] P. Drissen, Binding of trace elements in steel slags, in: H. Motz, P. Verhaag (Eds.), 5th European Slag Conference:Slag Products—Providing Sustainable Solutions for the Built Environment, EUROSLAG Publication, Luxembourg City, Luxembourg, 2007, pp. 217–227.

[20] J. Waligora, et al., Chemical and mineralogical characterisations of LD steel slag, in: H. Motz, P. Verhaag (Eds.), 5th European Slag Conference:Slag Products—Providing Sustainable Solutions for the Built Environment, EUROSLAG Publication, Luxembourg, Luxembourg, 2007, pp. 77–88.

[21] User guidelines for byproducts and secondary use materials in pavement construction. Steel slag. Publication Number: FHWA-RD-97–148. Accessed January 15, 2015, from http://www.fhwa.dot.gov/publications/research/infrastructure/structures/97148/ssa1.cfm.

[22] J. Emery, Steel slag utilization in asphalt mixes. National Slag Association Report MF 186-1. Accessed January 19, 2015, from http://www.nationalslag.org/sites/nationalslag/files/documents/nsa_186-1_steel_slag_utilization_in_asphalt_mixes.pdf.

[23] K. Kujala, Use of industrial co-products in civil engineering, in: H. Motz, S. Haimi, M. Mäkikyrö (Eds.), 4th European Slag Conference:Slags—Providing Solutions for Global Construction and Other Markets, EUROSLAG Publication, Oulu, Finland, 2005, pp. 63–70.

[24] D.D. Higgins, Soil stabilisation with ground granulated blastfurnace slag, in: J. Geiseler, M. Dean (Eds.), 3rd European Slag Conference: Proceedings Manufacturing and Processing of Iron and Steel Slags, EUROSLAG Publication, Keyworth, UK, 2002, pp. 53–62.

[25] N.O.D. Silveira, et al., ACERITA—Steel slag with reduced expansion potential, in: H. Motz, S. Haimi, M. Mäkikyrö (Eds.), 4th European Slag Conference:Slags—Providing Solutions for Global Construction and Other Markets, EUROSLAG Publication, Oulu, Finland, 2005, pp. 145–157.

[26] G. Wang, Y. Wang, Z. Gao, Use of steel slag as a granular material: Volume expansion prediction and usability criteria, J. Haz. Mat. 184 (2010) 555–560.

[27] D. Mombelli, et al., The efficiency of quartz addition on electric arc furnace (EAF) carbon steel slag stability,, J. Haz. Mat. 279 (2014) 586–596.

[28] A. Rozman, et al., Stabilization of ladle refining slags with borax, in: H. Motz, P. Verhaag (Eds.), 5th European Slag Conference:Slag Products—Providing Sustainable Solutions for the Built Environment, EUROSLAG Publication, Luxembourg, Luxembourg, 2007, pp. 137–145.

[29] Stainless steel—statistics. Accessed January 20, 2015, from http://www.worldstainless.org/statistics.

[30] L. Kriskova, et al., Effect of mechanical activation on the hydraulic properties of stainless steel slag, Cement Concr. Res. 42 (2012) 778–788.

[31] Y. Pontikes, et al., Options to prevent dicalcium silicate-driven disintegration of stainless steel slags, Arch. Metall. Mat. 55 (4) (2010) 1167–1172.

[32] J. Eriksson, B. Björkman, MgO modification of slag from stainless steelmaking, in: C. Pistorius (Ed.), VII International Conference on Molten Slags Fluxes and Salts, South African Institute of Mining and Metallurgy, Johannesburg, South Africa, 2004, pp. 455–459.

Environmental impact of ferrous slag usage in civil engineering

3

Slag is a waste material, and like any other waste disposed in landfills, it may affect the environment adversely. Therefore, in order to address the problem of existing landfills and prevent the need for new landfills, a current focus is finding new uses for slag. As civil engineering is an industry that depletes natural resources, the use of waste materials has increased greatly in recent years. From the sustainability point of view, the recovery of slag is positive from all perspectives, such as the following [1]:

1. *Economically:* Slag is cheaper than natural rock because it does not need to be mined, and if it is used, iron and steel companies do not have to pay disposal taxes.
2. *Environmentally:* Natural resources are spared and space in landfills is saved.
3. *Socially:* Slag is thought to have a low potential impact on human health and the environment.

According to Wang and Thompson [2], the comprehensive use of slag will have three significant benefits: substantial reduction in the dumping and stockpiling of this waste; protection of natural resources and energy requirements associated with obtaining natural materials; and finally, the possibility of altering or modifying properties of the basic materials to produce special engineering materials that can be used for specific applications.

There is no doubt that reusing slag reduces the deposited amounts of these types of waste materials, with the potential for saving materials from natural sources. However, before deciding on the possible uses of steel slag, as well as any other waste material, it is necessary to consider all the pros and cons for each particular area.

This chapter examines the impact of slag on the environment in terms of assessing its radioactivity and heavy metal contamination, which should be done before any potential use as a material in civil engineering or its disposal in a landfill.

3.1 Radioactivity of slag

The source of potential radioactivity in slag is metal scrap, which is used as a raw material and may have increased in radioactivity due to naturally occurring radionuclides or contamination from industrial or medical facilities. Worldwide studies describe a wide variety of findings regarding radiation contamination in steel scrap from the steelmaking process [3]. Such results support the idea of radiological control of waste that is used as construction material, because in addition to steel slag, fly ash may contain significant contamination with radionuclides. The most frequently present radionuclides in slag are ^{137}Cs, ^{60}Co, ^{226}Ra, ^{192}Ir, ^{241}Am, ^{232}Th, and ^{90}Sr [4], the increased levels of which may increase the exposure to ionising radiation of workers and the general population [5–8].

Characteristics and Uses of Steel Slag in Building Construction. http://dx.doi.org/10.1016/B978-0-08-100368-8.00003-8

Due to their long half-life, these radionuclides tend to be detected in slag before it is put to any further use. To be more precise, the activity concentration index may have a value of up to 0.5 mSv/y of limit values according to Radiation Protection Document 112 of the European Commission in cases when blast furnace slag is added to Portland cement [9]. Similarly, bricks made of slag may also be contaminated by radioisotopes [10]. Slag with high radium concentrations was used in some countries for decades as filling and insulating material in residential buildings, which in some cases caused annual exposure doses close to the occupational limits of 20 mSv/y [11,12].

There is specific concern when slag containing elevated concentrations of ^{226}Ra was used as a building material for schools and kindergartens [13]. The application of slag as roadfill may increase background radioactivity by several times [14]. In addition, the use of slag in roads and building materials may cause exposure to ionising radiation by workers manufacturing road materials and building products containing slag, workers constructing buildings and roads using slag, and the general population exposed in car parks, playgrounds, and houses constructed from materials containing slag [15]. The importance of investigations into and monitoring of the possible toxic effects of exposure to radiation on living systems may best be seen from the fact that certain road surfaces may have a slag content of up to 90% [15].

From the perspective of radiation protection, the strict control of radioactivity of steel slag in road construction and buildings is crucial, as the general population, including children, could be exposed continuously to annual doses of ionising radiation that may exceed 1mSv. According to some studies, radioactivity from slag does not reach levels that could cause health risks for the public in outdoor construction [7]. However, most countries prescribe reference levels for radioactive elements in building materials by applying activity indices or maximum permissible/recommended concentrations. The activity indices and maximally allowed or recommended concentrations in building materials intended for the utilisation of road structures in some European countries are given in Table 3.1.

Just as in the case of fly ash, the nature of the global market demands continuous control of raw materials, products, and waste. Contrary to historical industrial practices, production is no longer located close to the sources of raw material, but rather to the nearest and cheapest means of transportation. Thus, physicochemical characteristics originating from the different geological backgrounds of raw material (and, consequently, waste) have to be constantly monitored, as some continents are burdened with high natural radioactivity.

With regard to the current methods of radiation levels, online control for scrap metal may not be as reliable as on-site installed equipment for monitoring due to the shielding effect from uncontaminated scrap [3]. This is why additional measurements of final products containing slag are recommended.

The National Radiological Protection Board (NRPB) in the United Kingdom reported in 2003 that the release from all currently operating integrated steel production plants in the country was less than 10 μSv $y - 1$ [15], which was below the dose limit to the general population of 1000 μSv $y - 1$. The same report stated that radon

Table 3.1 Activity indices and maximum permissible/recommended concentrations of radioactive elements in building materials intended for use in roads in european countries [16–19]

Country	Activity index (Bq/kg)	Maximum permissible/ recommended concentrations
Germany	–	CRa = 200 Bq/kg
Finland	(CRa/700) + (CTh/500) + (CK/8000) + (CCs/5000) < 1	–
Latvia	(CRa + CTh)/300 < 1	CK < 2500 Bq/kg
Lithuania	(CRa/700) + (CTh/500) + (CK/8000) < 1	–
Norway	(CRa/300) + (CTh/200) + (CK/3000) < 1	CRa < 200 Bq/kg
Poland	0.0027·CRa + 0.0043·CTh + 0.00027 · CK < 1	CRa < 185 Bq/kg
Luxembourg	–	CK < 5000 Bq/kg, CTh < 250 Bq/kg and CRa < 350 Bq/kg
Slovakia	CRa + 1.25 x CTh + 0.086 x CK < 370	–
Croatia	(CRa/300) + (CTh/200) + (CK/3000) < 1	CK < 300 Bq/kg, CTh < 200 Bq/kg and CRa < 200 Bq/kg

Note: CTh, CRa, CK, and CCs are the activity concentration values of ^{232}Th, ^{226}Ra, ^{40}K, and ^{137}Cs in material, respectively.

concentrations in buildings constructed from materials containing slag that originated from radionuclides within the structure ranged from 7.0–10.8 Bq m^{-3}. As European Union (EU) guidelines [20] recommend that the amount of radium in building materials should be restricted to a level unlikely to exceed the design level for indoor radon introduced in EC recommendations (200 Bq m^{-3}), the estimated indoor radon activity concentrations were below this level.

The results of this NRPB report [15] also showed that when an emanation fraction of 0.7% for slag is assumed, the concentration of radon arising from the slag in the building materials is 0.6 Bq m^{-3}, and the total radon concentration arising inside buildings constructed using materials containing slag is 7.0 Bq m^{-3}.The estimated radon concentration inside buildings constructed without slag is 9.9 Bq m^{-3}. This implies that the use of slag as a component of building materials reduces the radon concentration inside the building. However, when an emanation of 5% is assumed for both slag and concrete, the estimated concentration of radon originating from the slag in the building materials is 4.4 Bq m^{-3}, and the total radon concentration inside buildings constructed using materials containing slag is estimated to be 10.8 Bq m^{-3}. In conclusion, the use of slag increases radon concentrations with respect to the use of standard building materials. The total estimated exposure level to a member of the public living in a house constructed using slag is 1.35 mSv y^{-1}. The estimated dose from living in a house constructed with standard building materials is 1.27 mSv y^{-1}. When an emanation of 0.7% is used, the reduction in the radon concentration in the interior air leads to a reduction in the radon inhalation dose and a dose that is lower overall by 0.12 mSv y^{-1} than that for standard building materials. It could be concluded that the use of slag in building materials does not have a significant radiological impact [15].

3.2 Environmental impact of cement replacement by slag

The production of cement releases greenhouse gas emissions both directly and indirectly. The heating of limestone releases CO_2 directly, while the burning of fossil fuels to heat the kiln emits CO_2 indirectly.

Figure 3.1 shows how cement replacement by ground granulated blast furnace slag influences the total CO_2 emissions per ton of cement, while Figure 3.2 shows the influence of slag proportion in cement on CO_2 caused by one of the three selected factors in the cement-making process [21]. A higher proportion of slag in cement can significantly reduce total CO_2 emissions. At the same time, the proportion of CO_2 emitted due to the calcination process, as well as the proportion of CO_2 caused by thermal energy production, will decrease with an increase in slag proportion, as will the proportion of CO_2 caused by electrical energy production.

By replacing 45% of ordinary Portland cement with blast furnace slag, a 37% reduction in total CO_2 emissions will be achieved [22]. The total energy costs (thermal and electrical) make up about 30%–40% of the total production costs of cement [23]. This is why efficient energy utilisation has always been a priority in the cement industry. Efforts to reduce demands by using more efficient equipment and substituting fuels and raw materials to lower production costs have been undergone in recent years. Figures 3.3 and 3.4 show how the cement replacement by ground granulated blast furnace slag influences the total energy, as well as the electrical-to-thermal energy demand ratio needed for cement production. On the one hand, total energy use decreases with

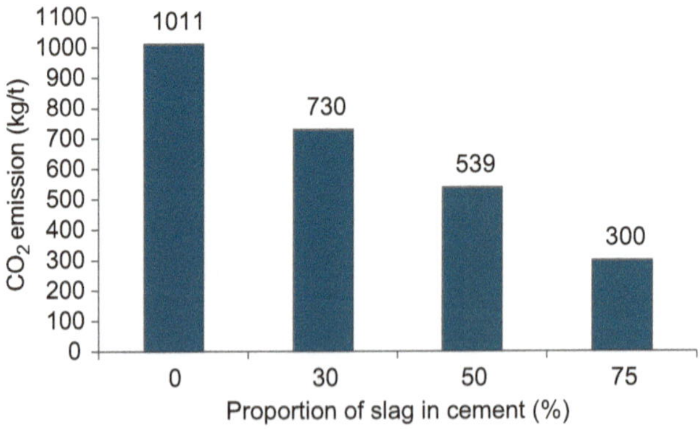

Figure 3.1 Influence of cement replacement by ground granulated blast furnace slag on total CO_2 emissions per ton of cement [21]

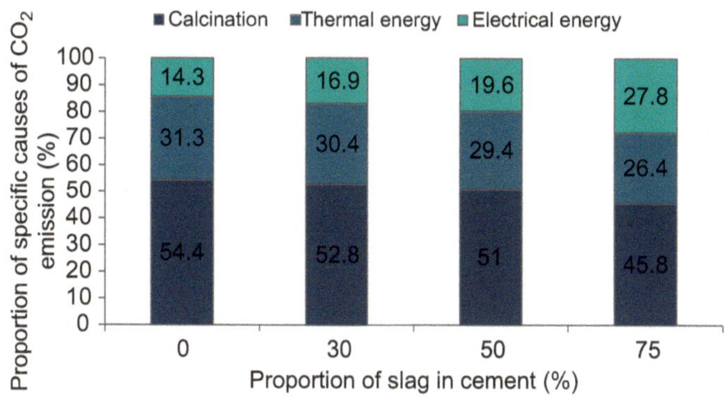

Figure 3.2 Influence of slag proportion in cement on CO_2 emissions resulting from specified causes in the cement-making process [21]

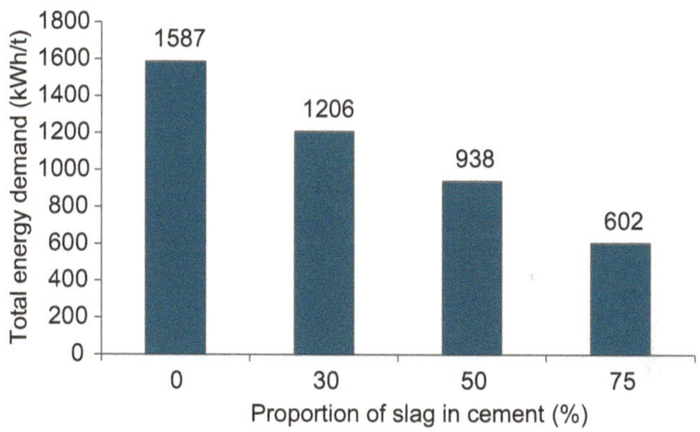

Figure 3.3 Influence of the slag proportion in cement on total energy demand in the cement-making process [21]

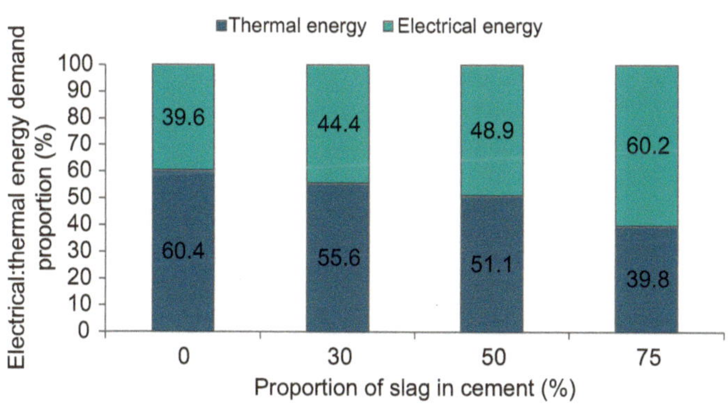

Figure 3.4 Influence of slag proportion in cement on the electrical-to-thermal energy demand proportion in the cement-making process [21]

Table 3.2 **Typical energy consumption for binder production**

Binder type	Energy consumption (MJ/t) [24]	Energy consumption (kWh/t)
Portland cement	5000	1389
Gypsum	5000	1389
Pulverised fuel ash	25	7,0
Granulated blast furnace slag	40	11,0

increasing slag contents, but on the other hand, the proportion of electrical energy increases. This is caused by the decreased amount of burnt clinker, the higher fineness of slag cement, and the lower grindability of granulated, glassy, blast furnace slag.

Calculated with the data for the typical energy consumption for binder production given in Table 3.2, the savings in total energy demand achieved by replacement of Portland cement by granulated blast furnace slag would be even higher.

However, the technology that prepares steel slag for application in the cement industry may contribute to air pollution. Therefore, care should be exercised in the production of cement from steelmaking slag. To be more precise, a study conducted in the United States (south Camden, New York) showed that approximately 2%–13% of the total deposited particulate matter (PM) mass came from the preparation technologies during production of slag into raw material for the cement industry [25].

3.3 Slag landfill—acceptance criteria

Even when deposited at a landfill for future use, slag must comply with the requirements given in the Landfill Directive (1999/31/EC) [26] regarding environmental impact. This directive requires the separate disposal of waste to landfills for inert, nonhazardous, and hazardous wastes and set out the concept of waste acceptance criteria for the different classes of landfill. Council Decision Document 2003/33/EC [27] provided acceptance criteria to be applied to granular wastes by member states. These included limit values for inert waste, stable, nonreactive hazardous wastes, and hazardous wastes for organic matter, as well as limits on the leachability of a range of predominantly inorganic parameters. The latter were to be applied to eluate generated at a liquid-to-solid (L/S) ratio of 2 and/or 10 l/kg prepared according to the EN 12457 [28–31] series of compliance leaching tests for granular wastes (Table 3.3) and/or initial C_0 eluate concentrations from the upflow percolation test according to CEN/TS 14405:2004 [32]. Since slag is not in the category of organic matter, the requirements given in [27] regarding limit values for organic matter should not be applied. Table 3.4 gives the limit values for inert waste such as slag. In order to classify slag as inert waste, the concentrations of specific elements and components in eluate prepared by one of the methods given in Table 3.3 or by percolation testing should be within the limits given in Table 3.4.

Table 3.3 Compliance test for leaching of granular waste materials and sludge

	L/S Ratio	Grain size of slag (mm)	Number of steps
EN 12457-1	2	4	1
EN 12457-2	10	4	1
EN 12457-3	2 + 8	4	2
EN 12457-4	10	10	1

Table 3.4 Leaching limits for inert waste

Component	L/S = 2 (mg/kg)	L/S = 10 (mg/kg)	C_0 (Percolation) (mg/l)
As	0.1	0.5	0.06
Ba	7	20	4
Cd	0.03	0.04	0.02
Cr (total)	0.2	0.5	0.1
Cu	0,9	2	0,6
Hg	0.003	0.01	0.002
Mo	0.3	0.5	0.2
Ni	0.2	0.4	0.12
Pb	0.2	0.5	0.15
Sb	0.02	0.06	0.1
Se	0.06	0.1	0.04
Zn	2	4	1,2
Cl	550	800	460
F	4	10	2.5
SO_4	560 (*)	1000 (*)	1500
Phenol index	0.47	1	0.32
DOC (**)	240	500	160
TDS (***)	2500	4000	-

*If the waste does not meet these values for sulphate, it may still be considered to be compliant to the acceptance criteria if the leaching does not exceed either of the following values: 1500 mg/l as C_0 at L/S = 0.1 l/kg and 6000 mg/kg at L/S = 10l/kg. It will be necessary to use a percolation test to determine the limit value at L/S = 0.1 l/kg under initial equilibrium conditions, whereas the value at L/S = 10 l/kg may be determined either by a batch leaching test or by a percolation test under conditions approaching local equilibrium.
**If the waste does not meet these values for dissolved organic carbon (DOC) at its own pH value, it may alternatively be tested at L/S = 10 l/kg and a pH between 7.5 and 8.0. The waste may be considered as complying with the acceptance criteria for DOC at its own pH or a pH range of 7.5–8.0, if the result of this determination does not exceed 500 mg/kg.
***The values for total dissolved solids (TDS) can be used alternatively for the values for sulphate and chloride.

3.4 Environmental impact of utilisation of slag in road, railway, and hydraulic construction

In construction, slag is usually employed in road and railway construction, as well as in hydraulic construction. In road construction, slag is typically used as a supplementary

Table 3.5 Typical energy consumption for aggregate production

Binder type	Energy consumption (MJ/t) [24]	Energy consumption (kWh/t)
Recycled aggregate	25	7.0
Primary crushed rock	50	14.0
Steel slag aggregate	40	11.0

cementitious material for hydraulically bound layers, as a replacement for aggregate for unbound and hydraulically bound materials, or in bituminous mixtures and surface treatments. In railway construction, slag is used as ballast, while in hydraulic construction, slag can be used for riverbeds, riverbanks/embankments, and armor stones. In most of these structures, steel slag is very often used, primarily because of its high density.

The way slag is applied depends on its physical and chemical properties, which vary depending on the source. For use in unbound layers, slag aggregate has to fulfil the requirements of EN 13242:2002+A1:2007 [33]. If slag is used for asphalt pavement, its properties must be in accordance with EN 13043:2002 [34].

Table 3.5, where typical energy consumption for aggregates production is given, makes it obvious why slag aggregate is so frequently used.

Since environmental acceptability is a fundamental principle for sustainable recycling, apart from civil engineering properties, there is an environmental aspect of slag application that has to be taken into account. To be more precise, in the application of slag in road, railway, and hydraulic construction, the possible harmful consequences for the environment should be taken into account since these layers are in direct contact with the surrounding soil and pollution can very easily reach groundwater. The major risk of pollution comes from the possibility of leaching heavy metals, which is described in more detail in the next section.

3.4.1 Leaching of heavy metals

The classification of waste materials such as slag with regard to soil and groundwater protection used to be based on the total contaminant concentrations measured in solids. However, the total solid pollutant content in waste material rarely relates directly with total harmful element content released into the environment. The leaching potency in materials such as slag is currently characterised by aqueous leaching tests considering the amount of harmful constituents being released in specific scenarios [35,36]. Leaching tests are used to determine the availability and actual release of elements that can intrude and contaminate the groundwater and are usually performed in laboratories in a controlled environment [37]. However, the release of heavy metals is desorption-limited, and hence not a function of the total solid pollutant content in the waste material. So, currently integrated approaches for environmental assessment are based on the results of leaching tests, which combine mathematical release and fate modelling and include technical and geographical factors [38]. Also, the toxicity and bioavailability of metal contaminants to terrestrial organisms depend on soil properties such as soil pH (in soils with low pH, many metals are more mobile and available for uptake than in neutral or alkaline soils) and weather conditions (e.g., rain, humidity) [39].

The origin of raw materials and technological processes in the iron and steel industries define the type and contaminants of slag. Comparing 150 studies on slag application, Piatak et al. [40] concluded that ferrous slag commonly has an acid-neutralising capacity and does not readily release environmentally significant amounts of most trace elements. In contrast, nonferrous slag may generate acid and release slag-type-specific trace elements when weathered. The elemental composition of slag depends on the feedstock composition and the metallurgic process employed, including any additives used. Undiluted leachates from slag may be toxic for aquatic organisms due to their specific sensitivity to elements such as Cu, Cr, Hg, Al, and Zn. In 2001, Japanese regulation limited levels of fluorine in iron and steel slag for use on land and strict control when slag-induced infiltration water is discharged to public waters [41].

Based on bioindicators, air cooled blast furnace slag has been reported to be the most toxic of all types of slag [7]. However, according to the results of studies on concrete containing ground granulated blast furnace slag, no leakage of substances that could cause possible health risks was detected [42]. Concerning blast furnace slag, sulphate as a mineral component may occur in higher amounts, but all tests confirm that this slag may be used in all areas of road construction except in areas with standing water. In this case, special tests have to be performed due to the conditions of the specific place in question [36].

It is a well-known fact that steel slag contains trace amounts of potentially toxic elements such as Fe, Mn, Ti, Al, Ca, Mg, and Cr [43]. In 2006, the UK Environment Agency and the Waste & Resources Action Programme in the United Kingdom initiated a project to develop a quality protocol for steel slag, with the aim of determining whether the application of steel slag could present risks to the environment or human health. The results showed that an impact on groundwater and surface water quality from steel slag is unlikely [44]. Also, studying the release of slag elements into the water, they concluded that other than the rise pH value, which is caused by partial lime melting from slag and the closely related electrical conductivity, the excretion of other elements does not have a significant impact on the environment [36]. However, many studies are still focused on the leaching of heavy metals, especially Cr and V, which is of the highest concern regarding their possible health effects or bioaccumulation (which makes these two metals worthy of receiving special attention in this chapter). The potential toxicity and mobility of possibly harmful elements such as Cr and V depends on their total concentration and, more significantly, on their chemical form or speciation [45]. The possibility of Cr and V leaching from basic oxygen furnace (BOF) slag was studied by Chaurand et al. [45,46]. They concluded that the leaching of Cr, despite its considerable amount in BOF slag, is very small and harmless to the environment. On the other hand, these authors point out that V leaching is also significant in its most toxic chemical form—the pentavalent form. According to [47], vanadium release is higher than Cr due to its mobility via oxidation from V(III) to V (V). Vanadium leaching is more present in carbonated slag due to the dissolution of Ca-vanadate [48]. Similar to the case of Cr, acidity increases the leachability of V (V) [49,50].

Tossavainen et al. [51] investigated the separation of heavy metals from steel slag that was modified in two ways for comparison with the original slag: remelting and water-granulation (rapid cooling) and remelting and cooling in the crucible (semirapid cooling). The conclusion was that the solubility of Cr, Mo, and V are very low, and that the rapid cooling of slag does not reduce the leaching of Cr.

In addition to laboratory leaching tests, Bialucha et al. [52] investigated the environmental impact of steel slag embedded in the surface layer of unpaved rural road. Two sections were built; the first had a pavement composed of a 40-cm base layer of unbound natural stone and a 10-cm surface layer of unbound electric arc furnace (EAF) slag, while the other had two layers of the same thickness built only from the EAF slag. During the study, seepage water was collected in the middle, on the edges, and at a distance of 5 m from the edge of the road, in two layers at 70 cm and 1 m below the pavement surface. They concluded that there is a difference between the results of research conducted in the laboratory and on the test sections, and that it is difficult to correlate these data. In this study, the composition of filtered water was not significantly different with natural stone and with slag, while the level of leaching depended on the subbase characteristics on which a layer containing the slag was built.

In the case of steel slag applications in concrete production, very low leaching of Cr (VI) and Cr (III) is reported due to its immobilisation in the spinel phase [53]. According to [54], tests of possible environmental problems due to heavy metal leaching from concrete with steel slag aggregates showed minimal leaching and fulfilled the requirements of the relevant standards. Leaching tests performed in the experimental bituminous mixtures containing steel slag resulted in acceptable levels, except for total Cr (16.1g/L) and V (147.4g/L) [55]. If present, the leachability of hazardous and radioactive agents may be reduced by combining such materials with alkali-activated binders [56].

Table 3.6 shows the parameters and allowable values for assessing the validity of steel slag applications in road and railway construction and hydroengineering structures according to German law. Such verification, in accordance with the same legislation, is to be conducted twice a year.

Normally, the pH value and electrical conductivity of BOF slag leachates are a little higher than in EAF slag leachates because of the higher Ca content. On the other hand, leachates of EAF slag may exhibit higher Cr content [57]. Table 3.7 gives the limit values for pH, electrical conductivity, and Cr content for these two types of steel slag.

Table 3.6 Important parameters and leaching concentration for steel slag in German law [36]

Analysed property	Units	Requirement
pH	–	10–13
Electrical conductivity	mS/m	500
Chrome	mg/l	0,03

However, in the United States, some states such as Maryland do not allow the use of waste materials with an effluent pH higher than 8.5 [58], which is a level that simultaneously increases the risk of heavy metal leaching.

In Europe, estimations of leaching are performed using the standardised leaching test, EN 12457 (1-4), which consists of a 24-h leaching sequence. EN 12457 consists of a four–part batch leaching procedures in which particle size–reduced material is extracted with deionized or demineralized water at room temperature. Parts 1, 2 and 4 of the standard are single–batch extractions at different L/S ratios and particle size requirements. Part 3 consists of a two–step test with re–extraction of the recovered material from the first L/S 2.0 mL/g extraction in a second extraction step at L/S of 8.0 mL/g (cumulatively the L/S is equal to 10 mL/g). However, this test does not assess the long-term effects caused by atmospheric conditions and wear of structures in which slag is incorporated.

The major source of possible contamination of food by toxic metals from slag used in road construction is the bioaccumulation of metals in plants, most frequently in their roots, which are known to have a high intake of heavy metals (such as carrots). In the case of plants, the highest bioaccumulation of V is reported to be in lucerne (alfalfa) [59]. As some plants are applied during phytoremediation, as they bioaccumulate heavy metals [60–62], it could be concluded that as a precautionary approach, such plant species should not be cultivated for human consumption next to roads and highways, as contamination by heavy metals may have origin both from traffic and from road construction materials.

Blast furnace and steel slag have also been used as liming materials and fertilisers for decades. Even though it is not a part of slag usage in civil engineering projects, it should be mentioned that the reduction in uptake of heavy metals from soil enriched with slag in plants has been reported [63–65]. In [10, 66], it was shown that by increasing soil pH, steel slag reduced bioavailability and heavy metal uptake (Cd, Zn, Cu, Pb) in rice by 60% in multimetal contaminated acidic soils. Recently, it has also been shown that steel slag, after being technologically processed, can be used as a fertiliser in seawater, as it can provide iron to seaweed [67]. In addition to releasing elements into the water, the effect of the steel slag used as armor stone on settlement of invertebrates has been monitored [36]. This study also showed that there is no difference between the effect of an ordinary natural rock and slag.

The Registration, Evaluation, and Authorisation of Chemicals (REACH) Working Committee developed a directive and safety data sheets for all types of iron and steel

Table 3.7 Important parameters and leaching concentrations for two types of steel slag [36]

	pH value	Electrical conductivity (mS/m)	Chromium content (mg/l)
BOF slag	≤ 11.5	≤ 100	≤ 0.02
EAF slag	≤ 11	≤ 80	≤ 0.03

slag. Following REACH's definition, most of the slag produced by the iron and steel industry throughout Europe will be registered as Substances of Unknown or Variable Composition, Complex Reaction Products, or Biological Materials (UVCB substances) and are to be used in applications such as the production of cements and concrete products.

3.4.1.1 Vanadium

Vanadium expresses several toxicological effects on living organisms, such as disturbances of the cardiovascular, respiratory, and nervous systems and metabolism. It is present both in animals and plants at very low levels ranging from 0.5 to 2.0 μg/g of dry weight but in cases of soil contaminated with high levels of V even in nonaccumulator plants, its levels may reach 144 μg/g. In plants, the majority of V mass remains in the root [68].

Cases of poisoning were described several times during the 20th century as slag was used widely as fertiliser [69]. Due to the trend of increased applications of V-containing waste, there is increasing concern regarding the health of the general population, because exposure to V in humans may result in health disturbances such as hypertension, dysrhythmia, systemic inflammation, hypercoagulation, various types of cancer, and bronchial hyperreactivity.

Due to extremely low levels of V in water and soil, measurements of its levels in water are not included in routine analyses in most countries and its upper permissible level is not measured and defined. Increased use of waste in construction and other industries, which may contribute to increased levels of V, will probably change this practice [70].

The toxicity of V leached from steel slag used in riverbank reinforcements on the freshwater aquatic organism hydroid *Cordylophora caspia* was reported in [71]. Vanadium leaching from steel slag with very low free lime content may cause an increase in V leaching above the limit values of some EU countries for both untreated and carbonated slag [72]. Accelerated steel slag carbonation at relatively low CO_2 pressure was used in order to improve environmental pH and the leaching properties of steel slag, with specific focus on V leaching. Carbonation experiments are performed in laboratory columns with steel slag under water-saturated and unsaturated conditions and temperatures between 5°C and 90°C.

Improvement of steel slag pH is not compatible with maintaining an acceptable leaching level of V. In the production of steel slag aggregates for construction, the di-Ca silicate phase contains a substantial level of V and is the major source of V leaching. In fresh steel slag, V leaching is limited by the presence of portlandite. Carbonation of the slag leads to the transformation of portlandite to calcite and the release of V. This is why future investigations should focus on secondary binding mechanisms for V in the carbonated slag mixture [72].

Vanadium has inhibitory effects on the population growth of the brackish water hydroid *C. caspia,* and the inhibition potential depends on water salinity (lowered salinity increases the inhibition potential). Asadpour et al. [73] investigated the bioaccumulation and effects of Ni and V on the mortality and growth of *Artemia urmiana* and *Artemia franciscana*, brine shrimps that live in saltwater lakes and ponds. After 24 h of exposure to Ni and V, results indicated that the mean length of animals on the 1st, 5th, 7th, and 11th day of life significantly decreases in comparison with control groups. Both species accumulate Ni and V in their bodies, and the results show that Ni is less toxic than V on *Artemia*.

In addition to exposure via leaching, it is important to stress that occupational exposure to V may occur from dust created by the grinding of slag. Inhalation of dust that contains several $\mu g/m^3$ of V will cause coughing, breathing problems, and eye irritation. It has a prolonged time of elimination from the organism (up to two weeks). Although there are data on health risks related with occupational exposure to V, there have been no such investigations of construction workers involved in technologies that use vanadium-containing steel slag [74].

3.4.1.2 Chromium

Chromium is a highly toxic metal whose hexavalent form is an epithelial irritant and carcinogen. It is also toxic for plants and animals. Contrary to the hexavalent form, trivalent Cr is a micronutrient that is important to sugar and lipid metabolism in humans.

The permissible levels of Cr differ. The Environmental Protection Agency (EPA) in the United States regulated the total Cr limit in drinking water to 0.1 mg/L. Following the report of the European Food Safety Agency from 2014 [75], there are no maximum levels for Cr in food. A parametric value of 50 μg Cr/L for total Cr in water intended for human consumption and a maximum limit of 50 μg Cr/L for total Cr are laid down in Council Directive 98/83/EC and in Commission Directive 2003/40/EC, respectively.

Chromium in soil is a mixture of the trivalent and hexavalent forms, and the levels of these two forms depend on a number of factors, such as the presence of other elements (Fe, V, or sulphides) and level of organic matter [76]. Chromium (VI) leaches at neutral pH and Cr (III) at the acidic pH range [77]. The process is not affected by carbonation [78].

As opposed to the Cr in slag, which is trivalent, almost all Cr in leachate is hexavalent. The oxidation of trivalent Cr to hexavalent depends on particle size as reaction occurs in areas near the particle surface [48].

Trivalent Cr in LD slag (Linz-Donawitz process, representing approximately 11% of steel production) is present in the trivalent form, which is less toxic and can accumulate in soil after long-term application. However, its uptake by plants is very low. Such low bioaccumulation is explained by a very strong fixation of trivalent chromium and other metals by slag to the soil, probably by the formation of sulphides [79]. Due to its high solubility, hexavalent Cr is also more bioavailable and toxic in lower concentrations than trivalent Cr. Despite this fact, recent regulations in Germany provide limits for levels of both trivalent and hexavalent Cr in fertilisers [80].

Among the plants with high uptake and bioaccumulation of Cr and used as food are wheat, lentils, peas, sunflowers, and Indian mustard (*Brassica* family), but in most plants, higher levels of Cr are also toxic [76, 81]. In plants, Cr mainly remains in the roots. Chromium (VI) bioaccumulates in higher concentrations than Cr (III). However, after Cr (VI) uptake, plants mostly convert it to the less toxic Cr (III) [82, 83]. The source of exposure of humans is consumption of plants grown on contaminated soil, such as fields next to roads [84]. In addition, crab and fish, also frequently consumed by humans, can bioaccumulate Cr [85].

In [86], leachability tests based on diffusion were performed in order to estimate the environmental impacts of asphalt mixed with slag. The results indicated that Cr in leachates from asphalt composites with the addition of slag was present almost solely in its hexavalent form. However, the concentrations were very low (below 25 μg/L) and did not represent an environmental burden.

3.5 Environmental impact of the utilisation of slag in wastewater treatment

Acid mine drainage (AMD) forms when sulphide minerals have been exposed to oxidising conditions during mining and other excavation activities, such as highway construction. In the presence of oxygen and water, sulphide minerals oxidise to form sulphate-rich and often metal-laden soil. AMD can be toxic to vegetation and can reduce the potability of water supplies. Various active and passive water treatment techniques have been developed during the last several decades to abate or control AMD. Active systems require the systematic and long-term addition of various alkaline chemicals, such as ammonia, hydrated lime, and caustic soda. These systems are often very effective at neutralising acidity. However, they are typically very expensive and require long-term commitments of personnel and maintenance. In contrast, passive treatment systems are less expensive and require very little maintenance. Passive treatment of AMD has largely relied on the dissolution of limestone to increase pH and alkalinity and precipitate metals. However, limestone is susceptible to armouring due to metal precipitation and has limited effectiveness in highly acidic or high flow discharges. A reliable, inexpensive alkaline source is needed for such situations. Laboratory and field studies indicate that highly alkaline steel slag may be a solution to this problem [87].

According to [79], blast furnace slag has a capacity to absorb phosphorus and several heavy metals, such as Cd, Cu, Pb, Zn, Al, and Ca. This property has great potential to assist in the development of new types of cheap, environmentally harmless, and effective filters for water cleaning, stabilisation materials, cover for landfills, and other purposes. Phosphorus removal was confirmed by using another type of slag, basic oxygen steel slag, and also showed the capacity of this slag in the removal of iron, a problematic metal in mine water [88].

3.6 Concluding remarks

Zero waste politics aims to reduce the environmental burden of the Earth from xenobiotics and save energy by recycling waste. The balance between the free-market economy and ecologically acceptable solutions is very unstable and sensitive. Human health risk and survival depend on this balance. Slag seems to be a good option as construction material, but like other similar waste types such as fly ash, its application has to be done in conjunction with controlling its environmental impact and dissemination of possibly harmful agents. The control of emissions of heavy metals or radiation from buildings or any kind of constructions containing slag according to the available analysis should be performed not only in laboratory conditions, but also periodically during their lifetime. This is because so far, we have no experience regarding the behaviour of such structures over several decades, nor have we seen their impact on the environment once they become construction waste.

In cases of uncontrolled conditions, large populations such as occupationally exposed populations, the general population, and ecosystems may be exposed to the emission of xenobiotics from slag. In addition, the wearing of roads built from steel slag may increase V concentrations in air particulate matter, which currently are not routinely measured. This is why the harmonisation between current technologies of waste usage and monitoring systems for air, soil, and water is of such major importance, as currently no groundwater standards are available for cobalt, tin, and vanadium. To achieve a controlled environment of slag usage, the Steel Slag Coalition (SSC) was founded in 1995. This association, consisting of the producers and processors of slag from ferrous metallurgy, independent and renowned chemical laboratories, and teams for risk assessment, is conducting research on the potentially negative impact of steel slag on humans and the environment [89]. The introduction of even low levels of highly toxic metals, such as hexavalent Cr and pentavalent V, demands close monitoring and further research, as soil is polyphasic in structure and its dynamics, fixation, and the release of elements and their uptake by the biosphere depend on a number of factors, such as pH, humidity, granular size, and temperature.

In the future, studies of the bioaccumulation of bioavailable metals in leachates from structures that contain slag should be encouraged, as extensive application of different types of wastes and slag has never been more important for human society.

References

[1] D.M. Proctor, et al., Physical and chemical characteristics of blast furnace, basic oxygen furnace, and electric arc furnace steel industry slags, Environ. Sci. Tech. 34 (8) (2000) 1576–1582.

[2] G. Wang, R. Thompson, Slag use in highway construction - the philosophy and technology of its utilization, Intl. J. Pave. Res. Tech. 4 (2) (2011) 97–103.

[3] T. Sofilic, D. Barisic, U. Sofilic, Monitoring of ^{137}Cs in electric arc steel making process, J. Radioanal. Nuclear Chem. 284 (3) (2010) 615–622.

[4] D. Neuschutz, et al., Inadvertent melting of radioactive sources in BOF or EAF: Distribution of nuclides, monitoring, prevention, ISIJ Intl. 45 (2) (2005) 288–295.

[5] A. Parmaksiz, et al., ^{226}Ra, ^{232}Th and ^{40}K radionuclides enhancement rate and dose assessment for residues of lignite-fired thermal power plants in Turkey, Rad. Protect. Dos. 147 (4) (2011) 548–554.

[6] T. Sofilić, et al., Radionuclides in metallurgical products and waste, Acta Metall. Slov. 10 (1) (2004) 29–35.

[7] L.A. Wendling, et al., Geochemical and ecotoxicological assessment of iron- and steel-making slags for potential use in environmental applications, Environ. Tox. Chem. 32 (11) (2013) 2602–2610.

[8] T. Sofilić, D. Barišić, U. Sofilić, Natural radioactivity in steel slag aggregate, Archives Metall. Mat. 56 (3) (2011) 627–634.

[9] S. Chinchon-Paya, et al., Radiological impact of cement, concrete and admixtures in Spain, Rad. Meas. 46 (8) (2011) 734–735.

[10] C. Zhou, et al., The environmental characteristics of usage of coal gangue in bricking-making: A case study at Huainan, China, Chemosphere. 95 (2013) 274–280.

[11] J. Somlai, et al., Radiation dose from coal slag used as building material in the Transdanubian region of Hungary, Rad. Protect. Dos. 118 (1) (2006) 82–87.

[12] T. Niewladomski, M. Olsezewska-Wasiolek, P. Wasiolek, Population exposure due to current building technology in Krakow, Poland, Rad. Protect. Dos. 1283 (1985) 293–296.

[13] C. Nemeth, J. Somlai, B. Kanyar, Estimation of external irradiation of children due to the use of coal-slag as building material in Tatabanya, Hungary, J. Environ. Radio. 51 (2000) 371–378.

[14] D.A. Krylov, G.P. Sidorova, Radioactivity of coal and ash-slag TPP wastes, Atom. Energy 114 (1) (2013) 56–60.

[15] National Radiation Protection Board (NRPB), Radiological impact on the UK population of industries which use or produce materials containing enhanced levels of naturally occurring radionuclides. Part II: The steel production industry, 2003, Transportation Research Board of the National Academies, National Radiation Protection Board: Chilton, UK.

[16] T. Sofilić, et al., Radioactivity of some building and raw materials used in Croatia, Polish J. Chem. Tech. 13 (3) (2011) 23–27.

[17] K. Kovler, Radiation of Building materials. [cited 2015 3.3.]; Accessed March 3, 2015, from http://tx.technion.ac.il/~cvrkost/Krina/sld001.htm.

[18] Radiation and Nuclear Safety Authority (STUK). GUIDE ST 12.2 2005, The radioactivity of building materials and ash, 2005, Accessed March 3, 2015, from http://www.finlex.fi/pdf/normit/23857-ST12-2e.pdf.

[19] Regulation on the conditions, manner, places and deadlines for systematic testing and monitoring of the type and activity of radionuclides in the air, soil, sea, rivers, lakes, ground waters, solid and liquid precipitation, drinking water, food and general use products and dwelling and working spaces, 2008, Official Gazette 60/2008.

[20] European Commission, Radiological Protection Principles concerning the Natural Radioactivity of Building Materials, Accessed March 15, 2015, from http://ec.europa.eu/energy/sites/ener/files/documents/112.pdf.

[21] A. Ehrenberg, CO_2 emissions and energy consumption of granulated blastfurnace slag, in: J. Geiseler, M. Dean (Eds.), 3rd European Slag Conference: Proceedings Manufacturing and Processing of Iron and Steel Slags, EUROSLAG Publication, Keyworth, UK, 2002, pp. 151–166.

[22] C. Hackländer-Woywadt, Grinding of granulated blastfurnace slag in Loesche vertical roller mills, in: H. Motz, P. Verhaag (Eds.), 5th European Slag Conference: Proceedings Slag Products—Providing Sustainable Solutions for the Built Environment, EUROSLAG Publication, Luxembourg, Luxembourg, 2007, pp. 181–192.

[23] A.M. Radwan, Different possible ways for saving energy in the cement production, Adv. Appl. Sci. Res. 3 (2) (2012) 1162–1174.

[24] H.L. Robinson, Maximasing the use of sustainable, green, recycled and low energy materials in high specification road construction, in: J. Geiseler, M. Dean (Eds.), 3rd European Slag Conference: Proceedings Manufacturing and Processing of Iron and Steel Slags, EUROSLAG Publication, Keyworth, UK, 2003, pp. 91–105.
[25] C.H. Yu, et al., Characterization of spatial impact of particles emitted from a cement material production facility on outdoor particle deposition in the surrounding community, J. Air Waste Manage. Assoc. 61 (10) (2011) 1015–1025.
[26] European Commission, Landfill Directive Regarding Environmental Impact, Landfill Directive (1999/31/EC), 1999.
[27] European Commission, Council Decision 2003/33/EC Establishing Criteria and Procedures for the Acceptance of Waste at Landfills Pursuant to Article 16 of and Annex II to Directive 1999/31/EC, 2003.
[28] European Committee for Standardization, EN 12457-1:2002 Characterisation of Waste. Leaching. Compliance Test for Leaching of Granular Waste Materials and Sludges. One-Stage Batch Test at a Liquid-to-Solid Ratio of 2 l/kg for Materials with High Solid Content and with Particle Size Below 4 mm (Without or With Size Reduction) 2002, CEN/TC 292: Brussels.
[29] European Committee for Standardization, EN 12457-2:2002: Characterisation of Waste. Leaching. Compliance Test for Leaching of Granular Waste Materials and Sludges. One-Stage Batch Test at a Liquid-to-Solid Ratio of 10 l/kg for Materials with Particle Size Below 4 mm (Without or With Size Reduction), 2002, CEN/TC 292: Brussels.
[30] European Committee for Standardization, EN 12457-3:2002 Characterisation of Waste. Leaching. Compliance Test for Leaching of Granular Waste Materials and Sludges. Two Stage Batch Test at a Liquid to Solid Ratio of 2 l/kg and 8 l/kg for Materials with a High Solid Content and with a Particle Size Below 4 mm (Without or With Size Reduction) 2002, CEN/TC 292: Brussels.
[31] European Committee for Standardization, EN 12457-4:2002 Characterisation of Waste. Leaching. Compliance Test for Leaching of Granular Waste Materials and Sludges. One-Stage Batch Test at a Liquid-to-Solid Ratio of 10 l/kg for Materials with Particle Size Below 10 mm (Without or With Size Reduction), 2002, CEN/TC 292: Brussels.
[32] European Committee for Standardization, 14405:2004 Characterization of Waste. Leaching Behaviour Tests. Upflow Percolation Test, 2004, CEN/TC 292: Bussels.
[33] European Committee for Standardization, EN 13242:2002+A1:2007 Aggregates for Unbound and Hydraulically Bound Materials for Use in Civil Engineering Work and Road Construction, 2007, CET/TC 154: Brussels.
[34] European Committee for Standardization, EN 13043:2002 Aggregates for Bituminous Mixtures and Surface Treatments for Roads, Airfields, and Other Trafficked Areas, 2002, CEN/TC 154: Brussels.
[35] C. Beyer, et al., Model-based prediction of long-term leaching of contaminants from secondary materials in road constructions and noise protection dams, Waste Manage. 29 (2009) 839–850.
[36] H. Motz, Production and use of air-cooled blastfurnace and steel slags, in: J. Geiseler, M. Dean (Eds.), 3rd European Slag Conference: Proceedings Manufacturing and Processing of Iron and Steel Slags, EUROSLAG Publication, Keyworth, UK, 2002, pp. 7–20.
[37] M. Tossavainen, E. Forssberg, The potential leachability from natural road construction materials, Sci. Total Environ. 239 (1999) 31–47.

[38] O. Schwab, et al., Beyond the material grave: Life cycle impact assessment of leaching from secondary materials in road and earth constructions, Waste Manage. 34 (10) (2014) 1884–1896.
[39] E.E.V. Chapman, G. Dave, J.D. Murimboh, A review of metal (Pb and Zn) sensitive and pH tolerant bioassay organisms for risk screening of metal-contaminated acidic soils, Environ. Poll. 179 (2013) 326–342.
[40] N.M. Piatak, M.B. Parsons, R.R. Seal. II, Characteristics and environmental aspects of slag: A review, Appl. Geochem. 57 (2014) 236–266.
[41] M. Nakagawa, The current state of the use of iron and steel slag products in Japan, in: H. Motz, P. Verhaag (Eds.), 5th European Slag Conference: Proceedings Slag Products—Providing Sustainable Solutions for the Built Environment, EUROSLAG Publication, Luxembourg, Luxembourg, 2007, pp. 167–217.
[42] Z. Giergiczny, A. Król, Immobilization of heavy metals (Pb, Cu, Cr, Zn, Cd, Mn) in the mineral additions containing concrete composites, J. Haz. Mat. 160 (2–3) (2008) 247–255.
[43] J.F.P. Gomes, C.G. Pinto, Leaching of heavy metals from steelmaking slags, Revista Metal. 42 (6) (2006) 409–416.
[44] Environment Agency, Steel slag quality protocol: Chemical risk assessment on BOF and EAF slags, Environment Agency, Bristol UK, 2013.
[45] P. Chaurand, et al., Environmental impacts of steel slag reused in road construction: A crystallographic and molecular (XANES) approach, J. Haz. Mat. 139 (3) (2007) 537–542.
[46] P. Chaurand, et al., Speciation of Cr and V within BOF steel slag reused in road constructions, J. Geochem. Explor. 88 (2006) 10–14.
[47] L. De Windt, P. Chaurand, J. Rose, Kinetics of steel slag leaching:batch tests and modeling, Waste Manage. 31 (2011) 225–235.
[48] M. Loncnar, et al., Investigation of the long-term leaching behaviour of EAF steel slag, in: H. Motz, P. Verhaag (Eds.), 6th European Slag Confrence: Proceedings Ferrous Slag—Resource Development for an Environmentally Sustainable World, EUROSLAG Publication, Madrid, Spain, 2010, 223–233.
[49] J. Macsik, A. Jacobsson, Leachability of V and Cr from LD-slag/Portland cement stabilized sulphide soil, Waste Manage. 16 (8) (1996) 699–709.
[50] V. Sjoberg, S. Karlsson, L. Sartz, Release of vanadium from LD-slag by exposure to ARD. IMWA 2010 "Mine Water and Innovative Thinking", 2010, Accessed March 29, 2015, from http://www.imwa.info/docs/imwa_2010/IMWA2010_13_06_sjoberg_s.pdf.
[51] M. Tossavainen, et al., Characteristics of steel slag under different cooling conditions, Waste Manage. 27 (2007) 1335–1344.
[52] R. Bialucha, R. Nicoll, T. Wetzel, Long-term behaviour of steel slag used for unbound rural roads, in: H. Motz, P. Verhaag (Eds.), 5th European Slag Conference: Proceedings Slag Products—Providing Sustainable Solutions for the Built Environment, EUROSLAG Publication, Luxembourg, Luxembourg, 2007, pp. 229–240.
[53] C.R. Panda, et al., Environmental and technical assessment of ferrochrome slag as concrete aggregate material, Const. Bldg. Mat. 49 (2013) 262–271.
[54] I. Papayianni, E. Anastasiou, Utilization of electric arc furnace steel slags in concrete products, in: H. Motz, P. Verhaag (Eds.), 6th European Slag Conference: Proceedings Ferrous Slag—Resource Development for an Environmentally Sustainable World, EUROSLAG Publication, Madrid, Spain, 2010, pp. 319–334.
[55] S. Sorlini, A. Sanzeni, L. Rondi, Reuse of steel slag in bituminous paving mixtures, J. Haz. Mat. 209–210 (2012) 84–91.
[56] C. Shi, A. Fernández-Jiménez, Stabilization/solidification of hazardous and radioactive wastes with alkali-activated cements, J. Haz. Mat. 137 (3) (2006) 1656–1663.

[57] T. Merkel, Successful utilisation of steel slags in Germany, in: J. Geiseler, H. Motz (Eds.), 2nd European Slag Conference: Proceedings Engineering of slags—A Scientific and Technological Challenge, EUROSLAG Publication, Dusseldorf, Germany, 2000, pp. 87–99.
[58] A.Y. Dayioglu, A.H. Aydilek, B. Cetin, Preventing swelling and decreasing alkalinity of steel slags used in highway infrastructures, J. Trans. Res. Board. 2401 (2014) 52–57.
[59] J. Yang, J., Vanadium uptake by alfalfa grown in V-Cd-contaminated soil by pot experiment, Bio. Trace Elem. Res. 124 (3) (2011) 787–795.
[60] D.I. Kaplan, et al., Vanadium: Toxicity and accumulation by beans, Water, Air, Soil Poll. 49 (1–2) (1990) 81–91.
[61] L. Agwaramgbo, et al., An evaluation of edible plant extracts for the phytoremediation of lead-contaminated water, J. Environ. Protect. 3 (2012) 722–730.
[62] S. Stasinos, et al., The bioaccumulation and physiological effects of heavy metals in carrots, onions and potatoes and dietary implications for Cr and Ni: A review, J. Food Sci. 79 (5) (2014) R765–R780.
[63] M. Rex, Environmental aspects of the use of iron and steel slags as agricultural lime, in: J. Geiseler, M. Dean (Eds.), 3rd European Slag Conference: Proceedings Manufacturing and Processing of Iron and Steel Slags, EUROSLAG Publication, Keyworth, UK, 2002, pp. 137–149.
[64] M. Rex, The use of BF, converter and ladle slags in European agriculture-benefits of risks? in: H. Motz, S. Haimi, M. Mäkikyrö (Eds.), 4th European Slag Conference: Proceedings Slags—Providing Solutions for Global Construction and Other Markets, EUROSLAG Publication, Oulu, Finland, 2005, pp. 51–62.
[65] M. Rex, Blastfurnace and steel slags as liming materials for sustainable agricultural production, in: J. Geiseler, H. Motz (Eds.), 2nd European Slag Conference: Proceedings Engineering of Slags: A Scientific and Technological Challenge, EUROSLAG Publication, Dusseldorf, Germany, 2000, pp. 137–149.
[66] H.H. Gu, et al., Mitigation effects of silicon rich amendments on heavy metal accumulation in rice (Oryza sativa L.) planted on multi-metal contaminated acidic soil, Chemosphere 83 (9) (2011) 1234–1240.
[67] M. Nakagawa, et al., Technology of construction seaweed beds by steel-making slag, in: H. Motz, P. Verhaag (Eds.), 6th European Slag Conference: Proceedings Ferrous Slag—Resource Development for an Environmentally Sustainable World, EUROSLAG Publication, Madrid, Spain 2010, pp. 89–102.
[68] B.G. Morrell, N.W. Lepp, D.A. Phipps, Vanadium uptake by higher plants: Some recent developments, Environ. Geochem. Health 8 (1) (1986) 14–28.
[69] A. Frank, et al., Vanadium poisoning of cattle with basic slag. Concetrations in tissues from poisoned animals and from a reference, slaughterhouse material, Sci. Total Environ. 181 (1996) 73–92.
[70] S.A. Grin, P.V. Kuznecov, I.V. Pitak, Impact of vanadium, Ekologya 6 (19) (2012) 9–12.
[71] U. Ringelband, Salinity dependence of vanadium toxicity against the brackish water hydroid Cordylophora caspia, Ecotox. Environ. Safety 48 (1) (2001) 18–26.
[72] A.V. Zomeren, et al., Changes in mineralogical and leaching properties of converter steel slag resulting from accelerated carbonation at low CO2 pressure, Waste Manage. 31 (11) (2011) 2236–2244.
[73] Y.A. Asadpour, P. Nejatkhah Manavi, M. Baniamam, Evaluating the bioaccumulation of nickel and vanadium and their effects on the growth of Artemia urmiana and A. franciscana, Iranian J. Fish. Sci. 12 (1) (2013) 183–192.
[74] T.I. Fortoul, et al., Overview of environmental and occupational vanadium exposure and associated health outcomes, J. Immunotox. 11 (1) (2013) 13–118.

[75] European Food Safety Agency, Scientific opinion on the risks to public health related to the presence of chromium in food and drinking water, Panel on Contaminants in the Food Chain (CONTAM), EFSA J. 12 (3) (2014).
[76] H. Oliveira, Chromium as an environmental pollutant: Insights on induced plant toxicity, J. Botany. 2012 (2012) 1–8.
[77] B. Dhala, et al., Chemical and microbial remediation of hexavalent chromium from contaminated soil and mining/metallurgical solid waste: A review, J. Haz. Mat. 250–251 (2013) 272–291.
[78] R. Baciocchi, et al., Effects of thin-film accelerated carbonation on steel slag leaching, J. Haz. Mat. 286 (2015) 369–378.
[79] L. Lind, Sorption of phosphorus and heavy metals on blast furnace slag, in: H. Motz, S. Haimi, M. Mäkikyrö (Eds.), 4th European Slag Conference: Proceedings Slags—Providing Solutions for Global Construction and Other Markets, EUROSLAG Publiation, Oulu, Finland, 2005, pp. 257–265.
[80] R. Bialucha, New regulations in the field of environmental affairs, in: H. Motz, S. Haimi, M. Mäkikyrö (Eds.), 4th European Slag Conference: Proceedings Slags—Providing Solutions for Global Construction and Other Markets, EUROSLAG Publication, Oulu, Finland, 2005, pp. 221–232.
[81] T.M. Chaudhry, et al., Phytoremediation-focusing on accumulator plants that remediate metal-contaminated soils, Australas. J. Ecotox 4 (1998) 37–51.
[82] H. Shanhandeh, L.R. Hossner, Plant screening for chromium phytoremediation, Intl. J. Phytoremed. 2 (1) (2000) 31–51.
[83] A. Zayed, et al., Chromium accumulation, translocation, and chemical speciation in vegetable crops, Planta 206 (1998) 293–299.
[84] B.B. Lind, A.M. Fällman, L.B. Larsson, Environmental impact of ferrochrome slag in road construction, Waste Manage. 21 (2001) 255–264.
[85] J.R. Peralta-Videa, et al., The biochemistry of environmental heavy metal uptake by plants: impplications for the food chain, Intl. J. Biochem. Cell Biol. 41 (2009) 1665–1677.
[86] R. Milačič, et al., Environmental impacts of asphalt mixes with electric arc furnace steel slag, J. Environ. Quality 40 (4) (2011) 1153–1161.
[87] P. Ziemkiewicz, J. Simmons, Use of steel slag as acid mine drainage, in: J. Geiseler, M. Dean (Eds.), 3d European Slag Conference: Proceedings Manufacturing and Processing of Iron and Steel Slags, EUROSLAG Publication, Keyworth, UK, 2002, pp. 167–182.
[88] L.I. Bowden, et al., The sustainable use of basic oxygen steel slag (BOF) to treat contaminated waters, in: H. Motz, S. Haimi, M. Mäkikyrö (Eds.), 4th European Slag Conference: Proceedings Slags—Providing Solutions for Global Construction and Other Markets, EUROSLAG Publication, Oulu, Finland, 2005, pp. 267–280.
[89] Steel Slag: A Premier Construction Aggregate. Accessed February 21, 2014, from http://www.acobrasil.org.br/siderurgiaemfoco/CCABrasil/NSA%20Steel%20Furnace%20Brochure.pdf.

Application of blast furnace slag in civil engineering: worldwide studies

4

Through the centuries, iron and steel slag was used for various purposes. The earliest report refers to Aristotle, who described the use of slag as a medicament in 350 B.C. [1]. However, the main field of application of this material was for construction purposes, and this remains the case today. Some of the first-known uses of blast furnace slag were in road building in the United States (since 1830), as a railroad ballast (since 1875), as a concrete aggregate (since 1880), and in bituminous surfaces (since 1900) [2]. A major milestone in the usage of blast furnace slag was German entrepreneur Emil Langen's discovery of the latent hydraulic properties, which took place in 1862 [3]. Its mass application as a cement additive began from that point. Half a century later, in 1909, the first German standard was published that permitted the use of slag in cement production, which legally confirmed the use of blast furnace slag in civil engineering [4]. Today, in accordance with current European cement regulations, blast furnace slag can be found in 14 types of cement.

From the early 20th century, production of slag increased along with the development of industry. According to EUROSLAG Association data [5], the production of blast furnace slag in Europe was 23.5 million tonnes in 2010 and nearly the same in 2012, with 23 million tonnes produced. These data are drawn from a study among EUROSLAG members, European steelworks and processing companies. In Figure 4.1, the production levels of different types of blast furnace slag, granulated and air-cooled, for the years 2010 and 2012 are presented. In Europe, the production of granulated blast furnace was predominant. The general applications of blast furnace slag, again according to EUROSLAG Association data, are presented in Figure 4.2 [5]. Most blast furnace slag is used in cement production or as an additive to concrete. Another important use of blast furnace slag usage is in road construction.

As is apparent from these graphs, blast furnace slag has been used in civil engineering for many years, and numerous scientific and professional papers have been published on the results of laboratory and field research. In the following sections of this chapter, some of the most common applications studied by researchers worldwide are presented, highlighting the wide range of possible applications for this valuable material.

4.1 Slag application in the cement industry

As described in Chapter 2, granulated blast furnace slag consists of the same major oxides as does Portland cement, which explains why its first application was in

Characteristics and Uses of Steel Slag in Building Construction. http://dx.doi.org/10.1016/B978-0-08-100368-8.00004-X

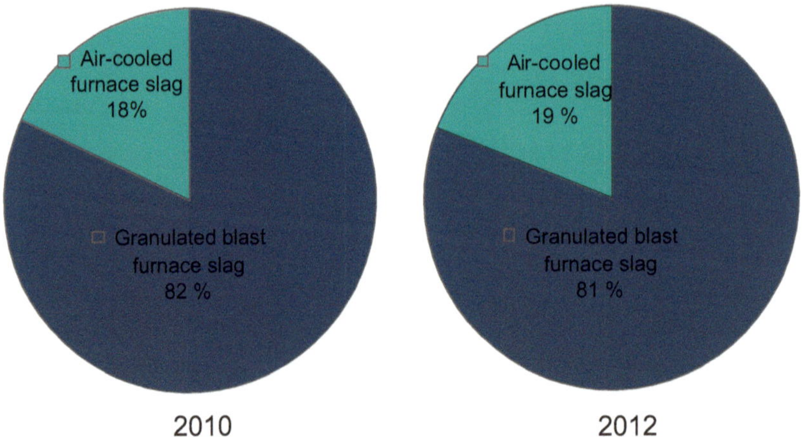

Figure 4.1 Production of blast furnace slag in europe [5]

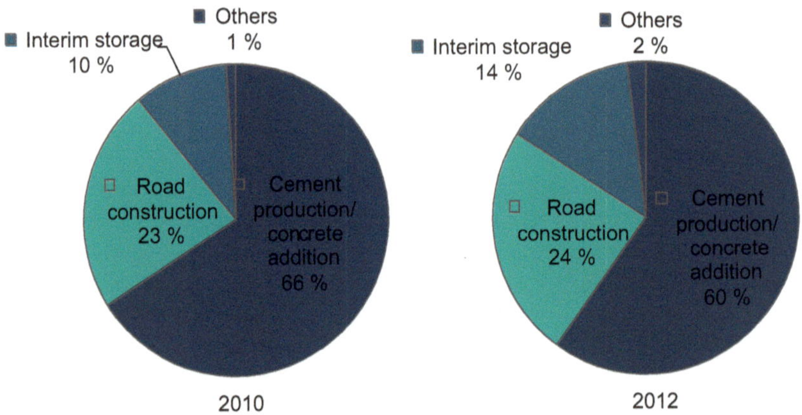

Figure 4.2 Use of blast furnace slag in europe [5]

the cement industry. However, the relative proportions of the oxides differ between Portland cement and this slag.

The first blast furnace slag cement was produced in Germany, in 1800, by mixing slag and lime [6]. Even before 1800, however, the properties of such cement were studied in France. Since the 1860s, it has been used in commercial production, and in the United States, slag cement production began in 1896 [7].

However, early attempts at using blast furnace slag in cement production were not successful. Poor performance of the initial slag cement was often due to poorly granulated slag and accelerators being added in large quantities without consideration of the long-term effects [7]. This even led to slag being prohibited in cement specifications. The reasons for abandoning production of this first type of cement in most countries were the inability of manufacturers to find suitable supplies of granulated slag, the slag's sensitivity to deterioration in storage, and its low strength in comparison to

Table 4.1 **Common cements containing blast furnace slag [16]**

Main types	Types of common cement		Main constituents (percentage by mass)			
			Clinker	Blast furnace slag	Other constituents	Minor additional constituents
CEM II	Portland-slag cement	CEM II/A-S	80–94	6–20	–	0–5
		CEM II/B-S	65–79	21–35	–	0–5
CEM III	Blast furnace cement	CEM III/A	35–64	36–65	–	0–5
		CEM III/B	20–34	66–80	–	0–5
		CEM III/C	5–19	81–95	–	0–5
CEM V	Composite cement	CEM V/A	40–64	18–30	18–30	0–5
		CEM V/B	20–38	31–49	31–49	0–5

Table 4.2 **Sulphate-resistant common cements containing blast furnace slag [16]**

Main types	Types of common cement		Main constituents (percentage by mass)		
			Clinker	Blast furnace slag	Minor additional constituents
CEM III	Sulphate-resistant blast furnace cement	CEM III/B-SR	20–34	66–80	0–5
		CEM III/C-SR	5–19	81–95	0–5

present-day Portland cement. Currently, a wealth of research exists on slag usage in cement production, processes have been developed for production, and the use of slag in cement is more developed. Possibly the most attractive feature of slag usage in cement production is the energy and financial savings in producing such a cement.

Today, blast furnace slag usage in the cement industry is known as *slag cement.* This term usually refers to either combinations of Portland cement and ground slag or to the ground slag alone. Within the current European cement standards, 14 listed cements may contain slag in different percentages by cement mass [8]. In Tables 4.1 and 4.2, the compositions and designations of common cements covered by

EN 197-1:2011 [16] can be seen. All these cements are mostly mixtures of ground Portland cement clinker and blast furnace slag. Granulated blast furnace slag should contain calcium oxide (CaO), magnesium oxide (MgO), and silicon dioxide (SiO_2), the sum of which should account for at least two-thirds of the mass. The remaining third contains aluminium oxide (Al_2O_3), with small amounts of other compounds. In addition, the ratio by mass (CaO + MgO):SiO_2 should exceed 1.0. Cement with blast furnace slag is usually intended to be equal to ordinary Portland cement (CEM I) and to meet the same requirements. Adding higher amounts of slag results in a higher sulphate resistance and lower heat of hydration properties. When producing such cement, slag is usually the hardest to grind, and it forms the coarsest portion, while the cement is ground the finest [9]. A potential problem is moisture, which occurs in blast furnace slag during the granulation process. The presence of moisture requires dewatering the blast furnace slag before using it as an additive or as a mineral admixture to Portland cement [10]. According to [11–13], this cement increases strength, reduces permeability, improves resistance to chemical attack, and inhibits rebar corrosion. In addition, blast furnace slag has a high resistance to freezing, thawing, chemicals, and seawater [14]. Therefore, it is recommended for concrete structures that require high durability. Concrete made with blast furnace slag cement sets more slowly than concrete made with ordinary Portland cement, depending on the amount of slag in the cementitious material. It also continues to gain strength over a longer period under production conditions. This results in less heat of hydration and lower increases in temperature, making avoiding of cold joints easier, but also possibly affecting construction schedules for which quick setting is required [13].

For a cement of extreme sulphate resistance, EN 197-1:2011 suggests that common cement contains blast furnace slag. Sulphate-resistant common cement is typically used for foundations because the presence of sulphates in the soil can lead to damage of ordinary cement [15].

The blast furnace cement (CEM III) in Table 4.1 is so-called low early-strength cement, indicated by *L*. Low early-strength CEM III cement conforming to the requirements in Table 4.2 can also be designated sulphate-resistant common cement. Common cement with blast furnace slag often has *LH* in its designation, which implies low heat of hydration.

In addition to EN 197-1:2011, European legislation includes five more types of cement containing blast furnace slag; four very low heat cements covered by EN 14216:2004 (Table 4.3) and a supersulphated cement covered by EN 15743:2010.

Very low heat cement has been found to be particularly useful in the grout for encapsulating nuclear wastes, and there may be other applications for which it is more important to avoid generating heat than to develop great strength (Table 4.4) [18].

Supersulphated cement is a mixture of ground granulated slag; calcium sulphate gypsum, which can be gypsum calcium sulphate dehydrate ($CaSO_4{\cdot}H_2O$), hemydrate ($CaSO_4{\cdot}1/2H_2O$), anhydrite ($CaSO_4$) or any mixture of these substances; and a small amount of Portland cement clinker [7, 9]. This cement has excellent resistance to aggressive agents (seawater, sulphates, chlorides, alkali hydroxides, and weak acids) and low heat of hydration. Compared with ordinary Portland cement, it is more susceptible to carbonation during storage and requires extra care during the initial curing period to

Table 4.3 **Very low heat cements containing blast furnace slag [17]**

Main types	Types of cement		Main constituents (percentage by mass)			Minor additional constituents
			Clinker	Blast furnace slag	Other constituents	
VLH III	Very low heat blast furnace cement	VLH III/B	20–34	66–80	–	0–5
		VLH III/C	5–19	81–95	–	0–5
VLH V	Very low heat composite cement	VLH V/A	40–64	18–30	18–30	0–5
		VLH III/B	20–38	31–50	31–50	0–5

Table 4.4 **Supersulphated cement [19]**

Main type	Type of cement		Main constituents (percentage by mass)		Additional constituents (percentage by mass)	
			Blast furnace Slag	Calcium sulphate	Clinker	Other
SSC	Supersulphated cement	SSC	≥75	≥5; ≤20	>0; ≤5	≥0; ≤5

keep the surface moist [7]. Due to its high resistance, this cement is generally used in seawater work, concrete pipes exposed to aggressive groundwater, or in chemical plants.

To justify the widespread application of blast furnace slag as an additive to Portland cement, manufacturers found significantly decreased consumption of natural raw materials, lower energy demands (grinding slag for cement replacement requires only 25% of the energy needed to manufacture Portland cement [10]), and specifically, lower CO_2 emissions [20]. Figure 4.3 presents the required energy to produce different types of cement. As can be seen, a possibility exists for a significant reduction in fuel consumption by utilizing pelletized slag. It requires only 15% of the energy of Portland cement clinker because this slag does not need to be dried [21].

For the production of cement or cementitious materials, any type of blast-furnace slag can be used, but the most common is granulated or glassy slag due to its ease in handling and more rapid reactions between the glass and other raw materials [7]. As cement is hydrated, calcium silicate hydrates (C-S-H) and calcium hydroxide (C-H) form. The main influence on the development of strength is C-S-H gel. When slag is added, its silicates react with C-H, forming additional C-S-H gel and creating denser,

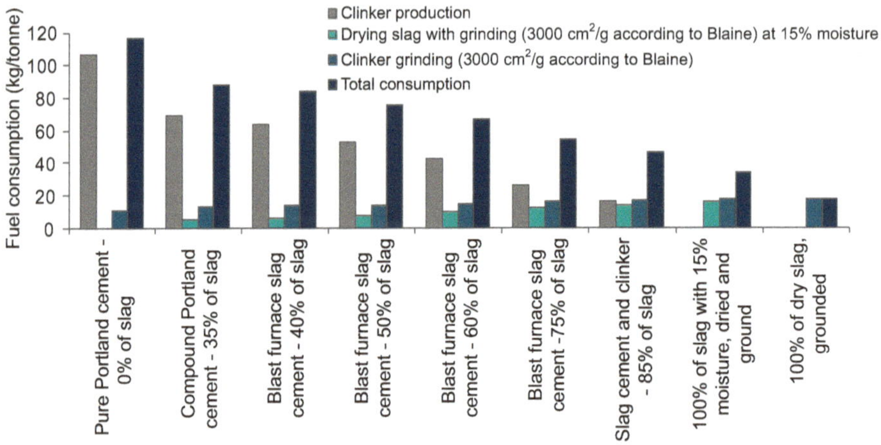

Figure 4.3 Fuel consumption for different types of cement production [21]

harder cementitious paste with increased strength [22]. However, the rate of hydraulic reactivity of granulated blast furnace slag is slower than for ordinary Portland cement; thus, slag with greater fineness is usually used [23].

Aside from being used as a cement additive, blast furnace slag can be added to concrete separately, thereby achieving an effect similar to concrete in its fresh and hardened state. In European legislation, the usage of blast furnace slag as an additive to concrete, mortar and grout is covered by EN 15167-1:2006.

4.2 Slag applied as an independent binder

Slag has recently been investigated as an independent binder, where its binding properties are enhanced by alkali activators. In this case, slag is called *alkali-activated slag* or *alkali-activated slag cement.* In general, alkali-activated cement is a system in which an alkaline activator promotes one or more reactions in a natural or artificial inorganic solid, generating a material with cementitious characteristics. One such material is alkali-activated slag cement, which results from mixing granulated blast furnace slag and alkaline substances.

Alkaline substances/activators are classified into six groups (with M being an alkaline):

- Caustic alkalis (MOH)
- Nonsilicate weak acid salts (M_2CO_3, M_2SO_3, M_3PO_4, MF, etc.)
- Silicates ($M_2O{\cdot}nSiO_2$)
- Aluminates ($M_2O{\cdot}nAl_2O_3$)
- Aluminosilicates ($M_2O{\cdot}nAl_2O_3$, SiO_2)
- Nonsilicate strong acid salts (M_2SO_4)

Of all possible activators, NaOH, Na_2CO_3, $Na_2O{\cdot}nSiO_2$, and Na_2SO_4 are the most widely available and economical [14]. Typical alkali-activated slag consists of ground granulated blast furnace slag with 3.5%–5.5% (in mass) of Na_2O added, usually as sodium hydroxide (NaOH) or water-glass [24].

The properties, workability, durability, and strength of alkali-activated slag are affected by factors such as the type and concentration of the alkaline activator, solution ratio, slag type and fineness, curing conditions, water/cement material ratio, activating solution/slag ratio, and the use of admixtures and fibres [25]. According to [26], the heat treatment has a significant accelerating effect on the strength development of slag paste. The same study concluded that sodium silicates provide the best activation, with the compressive strength of slag paste and mortar exceeding that of OPC (ordinary Portland cement) paste of the same water to binder (w/b) ratio. Increasing the amount of activators accelerates strength development and increases the 28-day strength of the paste [24].

Research on the alkali-activated slag has grown remarkably in recent decades because of numerous advantages such as fast development of strength, high resistance to chemical attack, and low heat release, as compared with normal Portland cement [24]. Furthermore, since the slag is an industrial by-product, the use of alkali-activated slag has many environmental advantages such as low energy cost, use of secondary raw materials, and low pollutant gas emissions. These advantages make it preferable to conventional binders in practice, and as a result, alkali-activated cement has found a variety of applications, such as transportation, industrial, agricultural, residential, mining, oil well cement, In addition, a major recent application is in waste management, including radioactive waste management and immobilization of toxic metals [27].

All the abovementioned research on alkali-activated slag as a binder has been conducted in controlled laboratory conditions by mixing slag with alkali solutions (i.e., alkaline substances added to water), but a great challenge that remains is to investigate the possibility of producing dry powder alkali-activated slag cement as a ready-to-use product that can be packed in bags and mixed with water at the construction site. Discovering a solution, in addition to publishing a European standard for alkali-activated binders, will definitely have a positive impact on the wider use of this binder within the European Union.

4.3 Slag applications in mortar

Another use of blast furnace slag in civil engineering was in mortar in the construction of the Empire State Building in the 1930s [28]. Due to its latent hydraulic properties, blast furnace slag is beneficial in mortar. The optimal use of slag in mortar has been observed to be 40% of the cement, which gives the concrete 19% more compressive strength and 25% more tensile strength than concrete made with ordinary Portland cement mortar after 180 days of curing. The corresponding increase in tensile strength is reported to be 25% [11].

Granulated blast furnace slag causes slower strength development under standard curing conditions, but at the same time, the ultimate strength is higher for the same water-cement ratio. This slower strength development is why blast furnace slag is not recommended for use in applications in which high early age strength is required, as previously mentioned. However, early age strength development can be influenced by higher temperatures, which is beneficial for the strength development for mortar containing a high level of blast furnace slag. Strength gain is much more rapid, even with an increase in curing temperature of only 10°C [29].

In addition to the usual applications of slag as a partial cement replacement and as an additive, recent research has investigated the use of slag as an alkali-activated binder in mortar. According to [30], higher compressive strength and tensile strength have been observed with higher curing temperatures. Another study [31] concluded that the highest strength for mortar can be achieved with 3.5% Na_2O as an activator at 60°C or with 6.5% Na_2O at 20°C, which means either with a higher concentration of alkali solution at a lower temperature or with a lower concentration of alkali solution at a higher temperature. These conditions resulted in strengths of at least 6 and 20 MPa at 1 and 28 days, respectively, which are quite high values for mortar. At 56 days, alkali-activated slag mortar showed more compressive strength than Portland cement mortar since the liquid/slag ratio does not influence the tensile strength of alkali-activated mortar [30]. In addition, higher drying shrinkage was observed in alkali-activated mortar than in Portland cement mortar [30]. Such conclusions on shrinkage have been confirmed [32]. According to [33], alkali-activated concrete and mortar have a lower total porosity and a larger fine pore structure than the respective Portland cement, which suggests a lower permeability and subsequently higher durability of alkali-activated mortar/concrete. In mortar containing alkali-activated slag binder with sodium silicate powder as an activator and river sand as the fine aggregate, fewer pores and a more compact structure could be achieved by increasing the slag content [32], which again supports the enhanced durability of such mortar.

Blast furnace slag can also be used as a fine aggregate substitute in mortar. Replacement of 25% to 75% of fine aggregate with blast furnace slag increases the compressive and split tensile strength of mortar and improves the mortar joint crushing strength and brick pull/adhesion strength [34]. In addition, partial replacement of fine aggregates with blast furnace slag can significantly enhance sulphate resistance and reduce rapid chloride permeability as a result of the improvement in the microstructure of the mixes [35]. However, increasing the amount of ground granulated blast furnace slag ratio used as a fine aggregate replacement (natural sand) results in a decreased flow of mortar [36]. Using a suitable concentration of chemicals and superplasticizer (0.5%–1%), the flow can be maintained and workability close to ordinary Portland cement mortar can be achieved.

4.4 Slag applications in concrete

Concrete is the most commonly and widely used civil engineering material, so blast furnace slag applications are widely investigated with the goal of sustainable development and environmental preservation. Slag can be used in concrete either as a binder or as an aggregate. As a binder, it can replace approximately 70–80 wt.% of Portland cement, with 9%–10% reduction of CO_2 emissions from concrete manufacturing for every 10% of Portland cement replaced [37]. Used as an aggregate, slag usually replaces fine grade aggregate and can improve the concrete's properties.

Slag usage in concrete is influenced by its fresh and hardened properties. Cement replacement by blast furnace slag results in more workable concrete, which is generally much easier to compact by vibration, resulting in less entrapped air [38]. Due to a higher fines content, low slag content concrete are easier to finish, but at high replacement levels and low ambient temperatures (< 15°C), setting times can be extended 1–2 h. In this case, extra care may be needed to guard against moisture loss. According to [39], 50% replacement of cement with blast furnace slag results in improved mechanical properties and durability of concrete. However, early age strengths are considerably lower for concrete that incorporates slag because of its low initial rate of hydration. Acceleration of hydration reaction can be achieved by using chemical activators (with solutions with pH > 12 creating alkali-activated binder, as described previously), by increasing its specific surface or by raising its temperature [40]. Consequently, the rate of cement replacement by blast furnace slag depends on its application, as well as on climate conditions.

The positive effect on concrete durability properties results from slag's latent hydraulic properties; secondary hydration reduces concrete porosity by blocking the capillary pores, resulting in lower permeability and diffusion coefficients than in ordinary Portland cement concrete [38]. This also makes the concrete microstructure denser, decreasing the penetration and diffusion of chloride ions and consequently achieving much longer transition periods and much lower chloride ion concentrations [41]. The pores become finer, and the change in mineralogy of the cement hydrates leads to a reduction in the mobility of chloride ions [42]. This enhanced durability of concrete with blast furnace slag favours its use in special fields of application, such as in constructing waterproof basements, dams, marine structures, bridges, telecommunication towers, and containment structures. A lower capillary porosity of concrete with blast furnace cement was reported in [43], which justified constructing a lot of catchment tanks in Germany with blast furnace cement. The beneficial effect of blast furnace slag on the alkali silicate reaction was reported in [44], and the effect has two parts: First, blast furnace slag cement has a denser pore structure than Portland cement, and this hampers different transport processes needed for alkali silica reaction. Second, blast furnace slag cement has a lower pH than Portland cement, which reduces the alkali silica reaction (a higher pH means that more hydroxide, which is needed to break down the silica networks, is present).

Low early strength development resulting from a low initial rate of hydration of slag can be beneficial due to the low heat of hydration, which can decrease the thermal cracking of mass concrete [28]. When concrete is exposed to a heavy tensile load or shrinkage, cracks occur, allowing the ingression of aggressive agents, such as chloride, carbonate, and sulphate, which can lead to the corrosion of reinforcement bars and degradation of concrete. In cement paste with a high percentage of blast furnace slag, self-healing of cracks can be achieved more efficiently than in ordinary Portland cement concrete, again contributing to durability [45].

Concrete durability largely depends on the possible corrosion of reinforcing bars in an aggressive environment, resulting in cracking and spalling of concrete. According to [42], the replacement of 40% of the cement by ground granulated blast furnace slag has no significant influence on corrosion rates of rebar in concrete, while with a replacement of 60% of the cement, the corrosion rate is significantly reduced. However, according to [41], the replacement of Type I cement (CEM I) by ground granulated blast furnace slag by 40% or more by binder weight can more than double the corrosion resistance of a steel bar. Sufficient concrete cover thickness and proper cement type can delay chloride-ion attacks in an aggressive environment and provide protection against corrosion. The addition of ground granulated blast furnace slag causes a reduction in the pH value but does not adversely affect the concrete corrosion resistance. A high percentage of cement can reduce calcium hydroxide compared with pore refinement, resulting in lower concrete resistance to carbonation [42]. In the presence of carbonation, scaling can occur under the combined stressors of freezing-thawing and deicing salt. Subsequently, other attacks on the structure are more likely to dramatically decrease the concrete's durability [28].

Concrete marine structures are probably the best way to test concrete durability due to the harsh environment conditions (e.g., salt and water presence and water splashing). Also, the ultimate test of the durability of structures is to measure their performance in practice. Research of 25- to 70-year-old marine construction made with blast furnace cement in the German part of the North Sea confirmed highly durable concrete; the main reason for problems were violations of the rules, such as high water/cement ratio or insufficient curing [46]. The higher capillary porosity in the carbonated surface layer of concrete made with blast furnace slag cement has been found to be the main reason for concrete scaling, while in laboratory conditions, the carbonation of blast furnace cement is much higher than in natural environments.

Besides mortar, alkali-activated slag can be used as an independent binder in concrete and with the same conclusions regarding the properties of concrete in its fresh and hardened state, as described in section 4.3. In addition, a good ductility of alkali-activated slag concrete is reported [47], which makes this concrete good to use where ductile designs are needed (i.e., in earthquake-prone areas).

As stated previously, the main disadvantage of using blast furnace slag in concrete is its reduced early strength development if used as a cement additive or replacement. However, if used as a fine aggregate, it will have a large surface area. Thus, more of it is available for hydration and pozzolanic reactions. Using ground granulated blast furnace slag as an ultrafine aggregate, workability and consistency of fresh concrete can be improved, yielding a consistent mix and higher compressive and flexural strength after 3 days; that is, early age strength compared to a control mix (with natural aggregate) can be achieved [48]. Self-compacting concrete with blast furnace slag used as the fine aggregate shows an increase in compressive strength at 365 days, higher autogenous shrinkage, self-desiccation generated by the slag hydration, and chemical shrinkage caused by slag reactivity [40]. Higher drying shrinkage as a result of slag's higher porosity increases with the percentage of aggregate replaced by slag.

Used as a coarse aggregate in concrete, blast furnace slag reduces concrete workability but increases compressive strength, while the splitting tensile strength and modulus of elasticity are similar to those of conventional concrete [49]. Also, concrete with coarse slag aggregate is more resistant to high temperatures; that is, this concrete has more residual compressive strength than conventional concrete.

Another characteristic of blast furnace slag concrete is lower chromium content, which is beneficial in terms of reduced skin irritation for workers who handle such concrete without any skin protection [28].

4.5 Slag application in soil stabilization

The latent hydraulic properties of blast furnace slag and alkaline activation are also useful in soil stabilization. According to [6], a combination of lime and granulated blast furnace slag is commonly used in South Africa and Australia for soil stabilization. The sulphate in the soil is known to be able to cause serious swelling in stabilized clay. This sulphate-related swelling is associated with the formation of ettringite. Granulated blast furnace slag cement is known to be resistant to this kind of expansion and swelling.

As used in soil stabilization, a combination of ground granulated blast furnace slag and lime results in slower early-rate strength development, ensuring prolonged time for construction work (particularly compaction and levelling after slag addition), the ability to self-heal in the case of early-life overloading damage, long-term increased strength, and inhibited deleterious swelling caused by sulphates or sulphides in soil [6].

Deep soil stabilization for both road construction purposes and building foundations may also be accomplished with alkali-activated blast furnace slag. Results of laboratory investigation indicate that cement-activated, ground granulated blast furnace slag has the potential for clay stabilization in deep stabilization when the aim is to achieve great pillar strength [50]. These results indicated that compressive strength increased with water content for stabilized clay, and the water/binder ratio decreased. With smaller amounts of binder, it is possible to achieve strengths of 300–600 kPa, and activated ground granulated blast furnace slag cement is a good binder when the aim of stabilization is high strength (in the range of 1000–3000 kPa).

Used as a soil stabilizer, slag can influence the environment, which has been discussed in more detail in Chapter 3.

4.6 Slag applications in road construction—pavement structure

As shown in Figure 4.2, about 24% of all blast furnace slag usage is in road construction. Mainly, it is used as an aggregate in asphalt and concrete mixtures or as an unbound base and subbase layer. This is mainly due to its rough surface and high porosity, which result in highly stable layers and practically no effect from water content variation, which is advantageous during construction.

The main advantages of blast furnace slag, if used as an aggregate in embankment or fill, are the high stability and friction angle (40–45 degrees), the ability to stabilize wet and soft underlying soil during construction time (particularly during early construction stages) and very little (or almost no) postcompaction settlement [10]. When used as an aggregate in embankment or fill, there is a possibility of an adverse, mainly aesthetic problem due to its chemical composition. Leachate from this material can be prone to discoloration and sulphurous odour (as discussed in more detail in Chapter 2) so an adequate drainage system is required. Also, in Chapter 3, the environmental impact of ferrous slag usage in civil engineering and all kinds of possible adverse environment impacts of slag are presented.

High stability and ability to "lock up" are also desirable when used in granular or unbound base layers. These properties provide good load transfer for pavements placed on soft, weak subgrade since angularity and high friction angle contribute to high bearing capacity with a California bearing ratio (CBR) greater than 100% [10]. Air-cooled blast furnace slag, even though is quite brittle and prone to breakdown under impact loading, has a hardness measured by the Mohs scale between 5 and 6 (similar to durable igneous rock).

Even though blast furnace slag is good aggregate material for use in road construction, combining it with different slag can result in material with enhanced properties. Combining blast furnace slag and ladle (LD) steel slag (produced during the Linz-Donawitz converter process) results in enhanced road construction material, with increased strength properties (uniaxial strength and CBR value) and freeze-thaw resistance, with the best results associated with a mixture of 50% granulated blast furnace slag and 50% LD steel slag [51].

Latent hydraulic properties of blast furnace slag can be used for the construction of stabilized or hydraulically bound base layers. For these materials, obtaining maximal dry density and optimal water content is very important during construction to obtain a well-compacted (and thus good load-bearing) base layer with less compactive energy. The addition of granulated blast furnace slag to a fly ash–cement mixture results in a well-graded mass, with increased maximal dry density and decreased optimal water content of the compacted mixture [52]. Slag also influences compressive strength increase and CBR value. This is attributed to the reaction of free calcium oxide, reactive silica, and alumina in slag during latent hydraulic reactions and additional formation of C-S-H gel.

In controlled low-strength materials, usually used as a backfill in lieu of compacted fill, ground granulated blast furnace slag increases slump flow and lowers the rate of bleeding [51]. However, the addition of slag does not significantly influence the workability and the water demand. It also increases compressive strength again as a result of its latent hydraulic properties. To keep compressive strength lower than the excavatability limit of around 2.1 MPa, slag content should be less than 50 kg/m^3.

Other than usage in bound or unbound base layers, blast furnace slag can be used as an aggregate in both asphalt and concrete surface pavement layers. In asphalt, it can be used as a coarse or fine aggregate. Due to its vesicular surface and absorption, if used as an aggregate in asphalt, it requires a higher binder content than conventional aggregate. For example, air-cooled blast furnace slag used as an aggregate in dense grade asphalt concrete demands asphalt cement content of about 8%, up to 3% more by weight of mix compared to conventional asphalt mix [10,53]. This characteristic of increased porosity compared to natural aggregate can also cause longer retention time for drying during hot mix asphalt manufacturing. With a slag shortage, if slag is used as aggregate in an asphalt mix, lack of consistency in the physical properties due to variability in the production process can cause problems during asphalt mix design and production, such as flushing due to high binder content or ravelling due to low binder content [10]. Also, resistance of air-cooled blast furnace slag to impact is fairly low, and the material can break down under heavy traffic [10].

Nevertheless, air-cooled blast furnace slag application yields numerous benefits in asphalt. A high affinity for asphalt cement as one of its negative properties also contributes to improved adhesive binding between the slag aggregate and binder, resulting in excellent stripping resistance. The aforementioned good surface characteristics, such as a hard, rough, highly angular, vesicular surface texture, result in high polish stone value (PSV), which also contributed good frictional resistance to a hot mix asphalt and surface treatment. So all advantages of slag usage in asphalt surface layers, like good friction resistance, good resistance to stripping, and fair wear resistance, and all stated disadvantages make slag asphalt mixes recommended for use on light traffic pavement and in areas where hard braking and acceleration must be taken into account, as for slurry seals [10].

Application of blast furnace slag in concrete pavement is possible according to the same principles mentioned in Section 4.3. The slag can be used as an aggregate in various types of concrete structures associated with road construction: pavements, precast and prestressed units, foundations, curbs, and gutters [53]. As in ordinary concrete, slag cement for use in concrete pavement decreases concrete workability and influences a decrease in total air content in the mix, resulting in less slump and increasing compressive strength, modulus of rupture, and modulus of elasticity compared to ordinary cement concrete, with less shrinkage [54]. In addition to ordinary concrete characteristics like compressive and flexural strength, concrete for pavement structure needs to be resistant to cyclical freezing and thawing, but also to the effects of deicing agents. Concrete's resistance to the action of deicing agents is highly influenced by air content. According to [55], slag cement concrete without an air-entraining admixture show significantly lower air content than ordinary Portland cement concrete, and for the same air content, the dosage of air-entraining admixture should be two times higher. Air entraining of concrete of as much as 5%–6% and high values of water/cement ratio result in high resistance of concrete to the action of deicing agents. This finding supports the use of slag cement in deicing salt-resistant concrete with a water/cement ratio higher than 0.45.

Concrete as a durable material is particularly suitable for airport pavement structures. High exhaust temperatures, high-velocity blasts, high-pressure tires, and fuel spillage from jet aircraft create one of the most aggressive environments. Besides the aforementioned data on reduced early strength but increased long-term strength, some data indicate that even for 7-day compressive strength, there is a decreasing trend in compressive strength; however, that trend changes at the 50% slag replacement level in cement paste [56]. In particular, that replacement level is stated to be the best suited to achieving the highest 28-day compressive strength of concrete, which can result in reducing the pavement thickness regardless of the design aircraft version (load) and subgrade reaction. However, pavement incorporating 70% blast furnace slag is more sustainable and eco-efficient in terms of fuel requirements and greenhouse gas emissions.

References

[1] J. Geiseler, I. Vaittinen, The status of processed slags from iron and steel making, in: J. Geiseler, M. Dean (Eds.), 3rd European Slag Conference: Proceedings Manufacturing and Processing of Iron and Steel Slags, EUROSLAG Publication, Keyworth, UK, 2002, pp. 37–42.

[2] D.W. Lewis, Properties and uses of iron and steel slags. Accessed January 12, 2015, from www.nationalslag.org/sites/nationalslag/files/documents/nsa_182-6_properties_and_uses_slag.pdf.

[3] H. Motz, Production and use of air-cooled blastfurnace and steel slags, in: J. Geiseler, M. Dean (Eds.), 3rd European Slag Conference: Proceedings Manufacturing and Processing of Iron and Steel Slags, EUROSLAG Publication, Keyworth, UK, 2002, pp. 7–20.

[4] D. Padovani, B. Corcoran, Improved performance of granulated blastfurnace slag (GBS) cements with cement additives. Cemtech Conference 2004. Accessed February 6, 2015, from www.mapei.it/dam/Pdf/ConferencesImproved.pdf.

[5] EUROSLAG. Statistics. Accessed February 6, 2015, from http://www.euroslag.org/fileadmin/_media/images/statistics.

[6] D.D. Higgins, Soil stabilisation with ground granulated blastfurnace slag, in: J. Geiseler, M. Dean (Eds.), 3rd European Slag Conference: Proceedings Manufacturing and Processing of Iron and Steel Slags, EUROSLAG Publication, Keyworth, UK, 2002, pp. 53–62.

[7] D.W. Lewis, History of slag cement. University of Alabama Slag Cement Seminar 1981. Accessed February 8, 2015, from http://www.nationalslag.org/sites/nationalslag/files/documents/nsa_181-6_history_of_slag_cements.pdf.

[8] EUROSLAG. Properties. Accessed January 15, 2015, from http://www.euroslag.com/products/properties/.

[9] D.W. Lewis, Cementitious applications of ground iron blast furnace slag, 1984, ACI Convention. Accessed February 10, 2015, from http://www.nationalslag.org/sites/nationalslag/files/documents/nsa_184-5_blast_furnace_slag_as_cement.pdf.

[10] User guidelines for byproducts and secondary use materials in pavement construction. Blast furnace slag. Accessed January 18, 2015, from http://www.fhwa.dot.gov/publications/research/infrastructure/pavements/97148/bfs1.cfm.

[11] M. Islam, et al., Strength behavior of mortar using slag as partial replacement of cement, The annual technical journal of the Military Institute of Science and Technology 3 (2011) 71–79. Accessed January 18, 2015, http://library.mist.ac.bd:8080/handle/123456789/207?show=full.

[12] R.M. Ferreira, et al., Blast-furnace slag cements for concrete durability in marine environment, in: CONSEC 04, 4th International Conference on Concrete under Severe Conditions, Environment and Loading, Seoul, Korea, 2004, pp. 109–116.

[13] D.S. Nandagawali, N.R. Dhamge, Study of blast furnace slag for improving mechanical property of concrete, Int. J. Eng. Sci. Inn. Tech. 3 (4) (2014) 473–477.

[14] H.-S. Kim, et al., Activation of ground granulated blast furnace slag cement by calcined alunite, Mat. Trans. 52 (2) (2011) 210–218.

[15] P. Bhatt, T.J. MacGinley, B.S. Choo, Reinforced Concrete Design to Eurocodes: Design Theory and Examples. 4th ed., CRC Press, Taylor & Francis Group, Boca Raton, USA, 2014.

[16] European Committee for Standardization, EN 197–1: 2011: Cement, Composition, Specifications, and Conformity Criteria for Common Cements, CEN/TC 51: Brussels, 2011.

[17] European Committee for Standardization, EN 14216: 2004: Cement, Composition, Specifications, and Conformity Criteria for Very Low Heat Special Cements, CEN/TC 51: Brussels, 2004.

[18] P.D. Bust, P.T. McCabe, Contemporary ergonomics 2005. Proceedings of the International Conference on Contemporary Ergonomics (CE2005), CRC Press, Hatfield, UK, 2005.

[19] European Committee for Standardization, EN 15743: 2010 Supersulfated cement, Composition, Specifications, and Conformity Criteria, CEN/TC 51: Brussels, 2010.

[20] A. Ehrenberg, CO2 emissions and energy consumption of granulated blastfurnace slag, in: J. Geiseler, M. Dean (Eds.), 3rd European Slag Conference: Proceedings Manufacturing and Processing of Iron and Steel Slags, EUROSLAG Publication, Keyworth, UK, 2002, pp. 151–166.

[21] R. Galibert, Slag cement, TC/MC and Executive Committee Meetings, 1981, Accessed February 8, 2015, from http://www.nationalslag.org/sites/nationalslag/files/documents/nsa_181-4_slag_cement.pdf.

[22] American Concrete Pavement Association, Slag cement and concrete pavements, 2003, Accessed February 8, 2015, from http://1204075.sites.myregisteredsite.com/Downloads/RT/RT4.03.pdf.

[23] B.S. Heaton, F.W. Down, J.J. Emery. Properties of ground granulated slags in cement blends. Accessed March 6, 2015, from www.nationalslagassoc.org.

[24] W. Chen, H.J.H. Brouwers, The hydration of slag, part 1: reaction models for alkali-activated slag, J. Mat. Sci. 42 (2) (2007) 428–443.

[25] W.A. Chaparro, D. Martínez Bastidas, J.H. Bautista Ruíz, Mechanical properties and absorption of chlorides in alkali activated slag concrete and exposed to carbonation. Revista Facultad de Ingeniería Universidad de Antioquia, 62, 2012, pp. 189–195.

[26] T. Bakharev, J.G. Sanjayan, Y.B. Cheng, Alkali activation of Australian slag cements, Cement Concr. Res. 29 (1999) 113–120.

[27] L. Chao, S. Henghu, L. Longtu, A review: The comparison between alkali-activated slag (Si+Ca) and metakaolin (Si+Al) cements, Cement Concr. Res. 40 (9) (2010) 1341–1349.

[28] F. Sajedi, et al., Relationships between compressive strength of cement-slag mortars under air and water curing regimes, Const. Bldg. Mat. 31 (2012) 188–196.

[29] S.J. Barnet, et al., Strength development of mortars containing ground granulated blast-furnace slag: Effect of curing temperature and determination of apparent activation energies, Cement Concr. Res. 36 (2006) 434–440.

[30] M.-C. Chi, J.-J. Chang, R. Huang, Strength and drying shrinkage of alkali-activated slag paste and mortar, Adv. Civil Eng. 2012 (2012) 1–7.

[31] O. Burciaga-Díaz, et al., Mortars of alkali-activated blast furnace slag with high aggregate:binder ratios, Const. Bldg. Mat. 44 (2013) 607–614.

[32] N.K. Lee, J.G. Jang, H.K. Lee, Shrinkage characteristics of alkali-activated fly ash/slag paste and mortar at early ages, Cement Concr. Comp. 53 (2014) 239–248.

[33] M. Palacios, F. Puertas, Effect of shrinkage-reducing admixtures on the properties of alkali-activated slag mortars and pastes, Cement Concr. Res. 37 (2007) 691–702.

[34] M. Nadeem, A.D. Pofale, Replacement of natural fine aggregate with granular slag—A waste industrial by-product in cement mortar application as an alternative construction material, Intl. J. Eng. Res. Appl. 2 (5) (2012) 1258–1264.

[35] T. Humam, R. Siddique, Properties of mortar incorporating iron slag, Leonardo J. Sci. (23) (2013) 53–60.

[36] M.C. Nataraja, et al., Use of granulated blast furnace slag as fine aggregate in cement mortar, Intl. J. Struct. Civil Eng. Res. 2 (2) (2013) 59–68.

[37] C.-J. Tsai, et al., Mechanical and cementitious characteristics of ground granulated blast furnace slag and basic oxygen furnace slag blended mortar, Mat. Des. 60 (2014) 267–273.

[38] R.D. Hooton, Canadian use of ground granulated blast-furnace slag as a supplementary cementing material for enhanced performance of concrete, Can. J. Civil Eng. 27 (2000) 754–760.

[39] M.L. Berndt, Properties of sustainable concrete containing fly ash, slag, and recycled concrete aggregate, Const. Bldg. Mat. 23 (2009) 2606–2613.

[40] M. Valcuende, et al., Shrinkage of self-compacting concrete made with blast furnace slag as fine aggregate, Const. Bldg. Mat. 76 (2015) 1–9.
[41] K.Y. Yeaua, E.K. Kim, An experimental study on corrosion resistance of concrete with ground granulate blast-furnace slag, Cement Concr. Res. 35 (2005) 1391–1399.
[42] H.-W. Song, V. Saraswathy, Studies on the corrosion resistance of reinforced steel in concrete with ground granulated blast-furnace slag—An overview, J. Haz. Mat. B138 (2006) 226–233.
[43] E. Lang, H.A. Brodersen, The performance of blast furnace slag cement, in: J. Geiseler, H. Motz (Eds.), 2nd European Slag Conference: Proceedings Engineering of Slag—A Scientific and Technological Challenge, EUROSLAG Publication, Dusseldorf, Germany, 2000, pp. 37–50.
[44] P. Crucq, Development of a Method to Measure the Mechanical Behavior of ASR Gels; Part I: Literature Study; Alkali-Silica Reactiocauses, Effects, and Prevention, 2005, Accessed March 6, 2015, from http://homepage.tudelft.nl/n89v3/LinkedDocuments/P_Qrucq_ASR_report_Literature.pdf.
[45] H. Huang, G. Ye, D. Damidot, Effect of blast furnace slag on self-healing of microcracks in cementitious materials, Cement Concr. Res. 60 (2014) 68–82.
[46] E. Lang, Freeze-thaw resistance of blastfurnace cements—Laboratory test and long-term experiences, in: H. Motz, S. Haimi, M. Mäkikyrö (Eds.), 4th European Slag Conference: Proceedings Slags—Providing Solutions for Global Construction and Other Markets, EUROSLAG Publication, Oulu, Finland, 2005, pp. 159–169.
[47] D.V.S.P. Rajesh, et al., Performance of alkali activated slag with various alkali activators, Intl. J. Inn. Res. Sci. Eng. Tech. 2 (2) (2013) 378–386.
[48] S. Teng, T.Y.D. Lim, B.S. Divsholi, Durability and mechanical properties of high-strength concrete incorporating ultrafine ground granulated blast-furnace slag, Const. Bldg. Mat. 40 (2013) 875–881.
[49] M.C. Pacheco, et al., Physical-mechanical properties and durability of concrete made with blast furnace and electric arc furnace slag as aggregate. in: H. Motz, P. Verhaag (Eds.), 6th European Slag Conference: Proceedings Ferrous Slag—Resource Development for an Environmentally Sustainable World, EUROSLAG Publication, Madrid, Spain, 2010, pp. 265–277.
[50] M. Mäkikyrö, GGBS as a binder in mix and deep stabilization, in: H. Motz, S. Haimi, M. Mäkikyrö (Eds.), 4th European Slag Conference: Proceedings Slags—Providing Solutions for Global Construction and Other Markets, EUROSLAG Publication, Oulu, Finland, 2005, pp. 99–108.
[51] A. Tikkakoski, K. Kujala, M. Mäkikyrö, Geotechnical properties of blast furnace slag and LD-steel slag mixtures for foar construction, in: H. Motz, S. Haimi, M. Mäkikyrö (Eds.), 4th European Slag Conference—Proceedings Slags: Providing Solutions for Global Construction and Other Markets, EUROSLAG Publication, Oulu, Finland, 2005, pp. 83–98.
[52] S.P. Singh, D.P. Tripathy, P.G. Ranjith, Performance evaluation of cement stabilized fly ash–GBFS mixes as a highway construction material, Waste Manage 28 (2008) 1331–1337.
[53] J. Emery, Slag utilization in pavement construction, in: W.W. Hotaling (Ed.), Extending aggregate resources, ASTM, Baltimore, USA, 1982, pp. 95–119.
[54] W.M. Hale, et al., Properties of concrete mixtures containing slag cement and fly ash for use in transportation structures, Const. Bldg. Mat. 22 (2008) 1990–2000.
[55] J. Deja, Freezing and de-icing salt resistance of blast furnace slag concretes, Cement Concr. Comp. 25 (2003) 357–361.
[56] A. Jamshidi, et al., Analysis of structural performance and sustainability of airport concrete pavements incorporating blast furnace slag, J. Cleaner Prod. 90 (2015) 195–210.

Applications of steel slag in civil engineering: worldwide research

5

Due to its volume instability, steel slag was not useful as a building material for a long time. However, with the growth of the steel industry, and the amount of slag produced as a by-product, meant that the disposal or recovery of slag needed to be addressed properly. The turning point in the usage of slag in construction was in 1880, when the application of steel slag as soil improver was confirmed [1]. Since then, the possibility of its application in many areas has been explored.

According to EUROSLAG Association data [2], the production of steel slag in Europe was 21.8 million tonnes in 2010, and nearly the same in 2012, with 21.4 million tonnes produced. These data come from a study of EUROSLAG members and European steelworks and processing companies. In Figure 5.1, the production levels of different types of steel slag in 2010 and 2012 are presented. The graphs show that in Europe, the production of basic oxygen furnace (BOF) steel slag is dominant as a result of the high share of BOF steel production in the total amount of steel produced. In 2013, the total crude steel production in the world was 1606.3 million tonnes, of which 1139.4 million tonnes was BOF slag and 451.3 million tonnes was electric arc furnace (EAF) slag [3].

The general applications of steel slag, again according to EUROSLAG Association data, are presented in Figure 5.2 [2]. Most of the steel slag produced is used in road construction. Final disposal and interim storage, over 30% of produced steel slag presents the opportunity for finding new areas of application and can be an encouraging area for new research into steel slag utilization.

Two of the most discussed adverse properties of steel slag when it is used in civil engineering are its volume instability and higher density compared to natural materials. There are several ways of dealing with this problem, the easiest of which is slag weathering over a specified period of time. As for higher density, that causes up to 21% higher lorry transport costs for steel slag than the cost of transporting the same amount of natural aggregate [4]. However, the use of slag reduces the amount of energy and emission that would otherwise be needed for natural aggregate extraction, crushing, screening, and cleaning, which compensates for this additional cost. In this chapter, this and other issues concerning steel slag utilization in civil engineering are addressed.

5.1 Use of slag as a raw material in clinker production

By definition, clinker is created by roasting limestone and clay. After the resulting material is ground and mixed with gypsum, cement is made. Replacing 20% a kiln feed with of slag allows for obtaining a kiln feed with steel slag allows for obtaining

Characteristics and Uses of Steel Slag in Building Construction. http://dx.doi.org/10.1016/B978-0-08-100368-8.00005-1

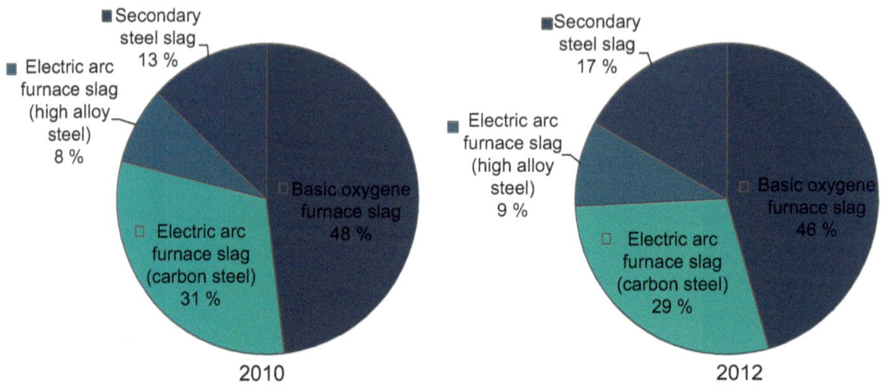

Figure 5.1 Production of steel slag in europe [2]

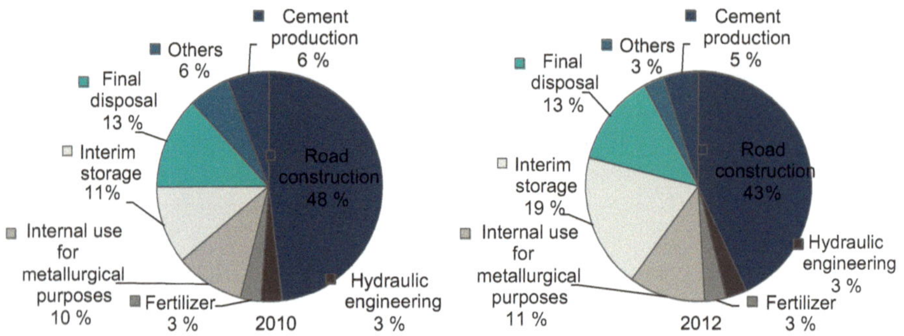

Figure 5.2 Uses of steel slag in europe [2]

better raw mix burnability in clinker production [5]. Also, the use of slag in clinker form makes the production process more environmentally responsible than conventional production methods. The presence of tricalcium silicate (C_3S), dicalcium silicate (C_2S), tetra calcium aluminate ferrite (C_4AF), and dicalcium ferrite (C_2F) provides steel slag with cementing properties. But the C_3S content is much less than in Portland cement, which is why the prevailing opinion is that steel slag used in a kiln feed produces inferior quality clinker. A comparative chemical analysis of Portland cement clinker obtained by a conventional charging furnace and clinker obtained by replacing the kiln feed with 10.5% of steel slag showed that the presence of steel slag in the furnace did not significantly influence the mineralogical composition of cement clinker and its microstructure [6]. By using steel slag in the production of cement clinker, cement is obtained with a slightly lower need for water to achieve standard consistency and delayed setting time. Despite the well-known high content of CaO and MgO within steel slag presenting potential volume instability problems, cement clinker with 6 wt.% and 14 wt.% of the fine fraction (< 160 μm) of steel slag from stainless steel production added to it seems to have a soundness of less than 1 mm [7]. This finding indicates that cement with steel slag clinker is not expansive above the admissible limits.

Compressive strength development steel slag clinker cement is similar to the development of compressive strength of ordinary cement clinker [6,7]. Based on these test results, it can be concluded that steel slag can be a high-quality raw material in a kiln feed during cement production.

The use of steel slag in clinker production can result in significantly reducing the environmental footprint of two industrial processes: cement and steel production. At the same time, the quality of the final products will not be compromised. Steel slag dumping can be reduced, but the greatest environmental impact of slag use in clinker production is the reduction of CO_2 emissions, which is achieved via several steps [7]. For slag utilization as a fine fraction of clinker production, this material requires no excavation and is very dry and fine; as a result, its preprocessing needs and costs are expected to be less than those of usual raw materials [7]. Mixtures with stainless steel slag most likely will require less energy during firing. Also, in view of the reduced limestone content in the raw blend, the energy required for limestone dissociation is reduced [7]. This limestone reduction can also result in lower CO_2 emissions during clinker production. Calculation of CO_2 emission reduction based on the chemical composition of the raw materials used in [7] showed that it is approximately 30 $kgCO_2/t$ when 6 wt.% slag is used and 65 $kgCO_2/t$ for 14 wt.% slag, compared to the reference sample (ordinary Portland cement). The potential hazardous impact of slag utilization in clinker production might be an increased level of Cr, which most likely will lead to higher Cr(VI) leaching levels. This problem can be addressed by introducing reducing agents such as ferrous sulphate, either the monohydrate ($FeSO_4{\cdot}H_2O$) or heptahydrate ($FeSO_4.7H_2O$) form, or stannous sulphate ($SnSO_4$).

5.2 Slag utilization as an aggregate in concrete

EAF slag contains a small proportion of amorphous silicon and a high proportion of iron oxide. It has little or no pozzolanic activity compared with blast furnace slag, which makes it unsuitable for use as a cement additive. The great hardness of steel slag has led to it being considered as an aggregate in concrete. Many studies have been conducted to explore this possibility.

In [8], steel slag was used to replace part of a fine aggregate fraction, in which silica sand (amounting to 15%, 30%, or 50%) or fine fraction was completely replaced by slag. In all the mixtures, the coarse aggregate was limestone. The influence that the percentage of slag had on the properties of fresh and hardened concrete in the strength classes of 25, 35, and 45 MPa was studied. As expected, due to the higher density of steel slag compared to natural aggregate, an increase in slag content increased the density of fresh concrete mix by 5%–20% over conventional concrete. This was stated to be one of the most important characteristics of steel slag aggregate concrete [9]. Also, as the share of slag in concrete increased, its consistency decreased.

As for compressive strength, replacing fine aggregate with slag by up to 50% achieved approximately the same or even higher values of compressive strength than the reference concrete, while concrete made of only slag aggregates had less compressive strength [8]. The positive effect of slag on concrete's compressive strength was more pronounced at the lower concrete strength class. The influence of slag on the tensile strength was more pronounced than its effect on compressive strength. Namely, increased slag content in the total aggregate mass increased the tensile strength of concrete. The positive effect of slag on the tensile strength was also more pronounced at the lower concrete strength class. Replacing natural aggregates by slag created 1.4 to 2.4 times higher tensile strength and up to 1.3 times greater compressive strength (with 15% slag aggregate). The cause of the decline in compressive strength with more than 50% slag aggregate was found to be due to slag´s fineness compared to sand. In particular, the increased amount of slag in concrete increases the total amount of the finely ground material. Therefore, a cement that will coat the grains is needed. By eliminating excessive amounts of fine particles in slag, there was a very positive effect on compressive strength at all maturing stages of concrete.

In [10], the effect of replacing a fraction of coarse aggregate in concrete with slag was studied. All the tested mixtures were made with the same amount of cement and with the same water-cement ratio. Compressive strength, ultrasound pulse velocity, absorption, and the corrosion of a mix with the ratio of slag in coarse fractions of 45%, 50%, 55%, 60%, and 65% of the total weight of the aggregate with the reference concrete with limestone (with a coarse-to-fine ratio of 60:40) were compared. Increasing the slag portion in the total mass of aggregate made the compressive and tensile strength of concrete increase as well. With the same coarse and fine aggregate ratio (60:40), slag concrete gained a slightly higher compressive strength and a slightly lower tensile strength compared to the reference concrete. The absorption and porosity of the mixture decrease as the slag portion increases, suggesting possible better durability properties of such concrete [10]. Ultrasound pulse velocity increased with the increase in slag portion, suggesting a greater modulus of elasticity of concrete with slag as aggregate in relation to the reference concrete.

A reduction in compressive strength and ultrasound pulse velocity after the exposure to thermal variations during 60 cycles was also observed. One cycle consisted of exposing the samples to 70°C for 8 h and to 25°C for 16 h. Compressive strength decrease during thermal variation was reduced with slag aggregate portion increase in mixture. The effect of thermal variation on the ultrasound pulse velocity (and thus on the modulus of elasticity of concrete) was less pronounced in the concrete that contained slag than in the reference concrete. Although the increase in absorption due to thermal variations was more pronounced in mixtures containing slag than in the reference concrete, absorption of concrete containing slag was certainly less after heat cycles than the absorption of reference concrete [10]. In the same paper, the possibility of an aggregate corrosive action on steel was also studied. A substantial increase in the corrosion resistance of steel built in concrete containing slag compared to the reference concrete was found, and steel corrosion was more delayed with the increase in slag aggregate.

When analysing the influence of steel slag aggregate on the mechanical characteristics of concrete (i.e., compressive and tensile strength), the interfacial transition zone (ITZ) between the aggregate particles and the cementitious matrix need to be considered. This ITZ is considered to be a weak zone of the concrete. In this region, the appearance of microbleeding around the aggregate particles, porosity, and some microstructural features depend on several factors, such as aggregate quality and size, the water-cement ratio, the binder, and the age of the mix [9]. Also, the morphology (i.e., quality) of ITZ plays an important role in concrete permeability and durability.

The presence of EAF steel slag in concrete creates a different ITZ morphology. In particular, a smaller and less hollow ITZ compared to natural rock aggregate concrete is a result of the slow migration of CaO from the core of the steel slag grains to its surface, resulting in the chemical evolution of CaO to calcium carbonate [9]. Due to a stronger ITZ within EAF steel slag concrete, shown by broken aggregate particles on the fracture surface after concrete mechanical rupture testing, improved concrete mechanical strength (compressive and tensile) results. However, since steel slag has a porous texture with multiple cracks and crevices, breakage under lower loads can be expected. In order to reduce the impact of this porous structure on strength characteristics in concrete, steel slag aggregate can be pretreated by polishing in a Los Angeles abrasion machine. This polished coarse aggregate seems to have an improved quality, with a smoother surface and resistance to microcrack emersion. This results in a higher surface-dry density and lower water-absorption rate compared to regular, untreated steel slag aggregate [11]. Also, polishing treatment of coarse slag aggregate has an impact on concrete fatigue failures. For concrete with untreated slag aggregate, fracturing of aggregates can be observed, and fatigue failure ends in the aggregate-fracture type. On the other hand, in a case of polished slag aggregate, most of the aggregates are sound and fatigue failure ends in the interface-peeling type between paste and aggregate [11].

The characteristics of polished slag aggregate result in improvements in concrete characteristics such as fatigue strength, strain, acoustic emissions, and hardness of the materials under compressive fatigue loads. The other way of improving slag aggregate quality (particularly the ITZ) is by an accelerated carbonation treatment of steel slag aggregate. This carbonation process is conducted in a carbonation reactor that has been sealed to 70°C and vacuumed to −0.3 MPa [12]. Then CO_2 was introduced into the reactor until the pressure reached 0.3 MPa. This process caused a change in the aggregate pore structure. It reduces the occurrence of pores with the diameter larger than 1 μm by 24.4% and increases the occurrence of pores with a diameter of less than 1 nm by 67.9% [12].

After carbonation, the steel slag aggregate water absorption rate decreased, while its impermeability properties increased. The expansion ratio of the steel slag is also reduced due to the reduction of free CaO content. For comparison of ITZ strength for concrete with natural and steel slag aggregate, strength of carbonated steel slag and natural aggregate was compared so as concrete compressive strength [12]. Although the crushing strength of carbonated slag aggregate was less than that of natural aggregate, compressive strength of slag concrete has improved. When the concrete sample cross section was subjected to compressive strength testing, cracks did not go through the middle of the aggregate but traveled around the grain its self. Thus, the ITZ of the carbonated steel slag aggregate concrete is stronger and more rugged than that of natural aggregate concrete.

Carbonation of slag aggregate also has a beneficial environmental impact—namely, reducing the leaching potential of certain elements. In analysis of the leaching characteristics of stainless steel slag aggregate, Ca and Si were found to be those elements most appreciably affected by carbonation due to the changes in the mineral phases responsible for solubility control of these elements [13]. The leachability of Cr, one of the most toxic elements in slag aggregate, appeared not to be significantly affected by carbonation, although Mo showed some reduction in leaching.

5.3 Slag utilization as mortar aggregate

In addition to being used as concrete aggregate, slag can be used as an aggregate in mortar for masonry. The limiting parameter in the application of slag in mortar, given the small amount of thickness of the mortar for masonry, could be its expansive nature. The volume changes of mortar containing slag were investigated and compared with the volume changes of mortar with sand [10]. These changes were measured in the samples cured for 7 days in a humid environment and exposed to drying in the laboratory at 25 ± 2°C and 50 ± 5% relative humidity. The results pointed to an increased shrinkage of samples with sand compared to samples containing slag. Also, the change in length of mortar samples with sand and those with slag during exposure to a humid environment for a period of 4 months was tracked, and a significantly greater expansion was observed in the slag samples.

The influence of grain size of the slag as aggregate in the mortar on compressive strength has been investigated [14]. Steel slag was used, which was first weathered for 5 months and then was kept moist for 15 days in the laboratory before the test in order to reduce the possibility of expansiveness. The result of this treatment was the complete absence of free lime, which was reflected in the almost complete absence of expansion. The conclusion of this study was that slag granulation has a significant effect on the materials' compressive strength, as well as on its water absorption. In particular, due to the increase in grain size, the grain-specific surface area is reduced, as well as the water required to achieve optimum moisture and absorption. With the increase in coarse fraction, increases in compressive strength occur as well.

5.4 Slag utilization in unbound base layers in pavement

Unbound base layers of pavement structures are mostly made of natural materials. Since these layers are substantially thicker than other layers (for instance, the average thickness of unbound base layer is of 20–45 cm, as opposed to 2–8 cm for the asphalt wearing layer), the use of alternative materials would contribute significantly to environmental conservation. Thus, it is not surprising that the first use of slag in road construction was for the purpose of building these layers. About 35 years ago, a few test sections were built in Germany with slag embedded in unbound base courses [15]. A comparison of layers containing steel slag and layers with natural aggregate in crushed gravel was performed. Layers containing slag achieved a higher bearing capacity immediately after compaction, which is a big advantage over natural aggregate.

Specifically, pavement layers are very sensitive to external influences during construction due to the movement of heavy machinery. The increase in strength is caused by carbonate hardening due to free calcium oxide. Increased strength also occurs due to the specific structure of slag grains, which favours the fixing of grains after installation and compaction and forms a solid, compact, and durable surface that can accommodate heavy traffic [16]. For this reason, this material also can be used for constructing roads on weak soils since good grain interconnectivity provides good load transfer to the formation soil. Also, the occurrence of heavy rain during construction does not affect the quality of the built layers, and slag inbuilt into unbound base layers does not pollute the environment due to leaching [15].

5.5 Slag utilization in stabilized base courses in cement

Due to the positive characteristics of slag observed through its application in unbounded base layers, this material is being explored for use in the construction of stabilized (hydraulically bound) base layers. At first, the properties of different types of metallurgical slag (including EAF slag) with 2% of cement and lime added were studied in order to determine the impact of binders on the strength of the mixture [17]. This small amount of binder cannot act as a binder by itself, but it does activate slag. Of all the tested slags, EAF slag is the only one that consistently showed the advantage of activation throughout every stage of hardening. Binder addition significantly accelerates the development of strength in all types of ferrous slag, which is especially pronounced during the first year of the material's life.

In Belgium, during the construction of storage plateaus, steel slag generated from the production of stainless steel was used as an aggregate in cement stabilized base layers [18]. The mix composition was as follows: 78% slag (grain size 0/20 mm), 18% sand, 4% cement, and optimum moisture content of 8%. After the first day of mixture placing, volume changes of 0.4% was observed and 1.8%–2.9% after 7 days (an average volume change of 2.3%). The value of volume change did not represent a major problem in the construction and use of the plateau. To demonstrate the suitability of slag for use in stabilized base layers, 28 days of compressive strength was determined under a curing regime of 20°C temperature and humidity of 90%. The measured compressive strength was 7 MPa.

Many authors cite the poor cementing and pozzolanic properties of steel slag [19–22]; therefore, new ways to improve that property have been found, and exploration in this area continues. For example, improvement of cementing properties of EAF slag can be achieved by a special way of processing it before use [21]. Specifically, if the EAF slag is subjected to a remelting process and then cooled by water, there are significant changes in its composition and structure. Laboratory tests have shown that this method can increase pozzolanic strength by up to four times [21]. However, in BOF slag, lack of pozzolanic reaction after 28 and 90 hardening days was observed, and the pozzolanic reaction cannot be achieved using only this type of slag [22]. Therefore, BOF slag cannot be applied as the main constituent of hydraulic binders in road construction, but its free oxides can be used to activate other materials, such as blast furnace slag. For example, a mixture of blast furnace and BOF slag (with ratio of 52.5% of blast furnace slag and 42.5% of BOF slag) is the ideal, with the addition of the activator in the amount of 5%. This mixture does not show any problems in terms of expansion, and it was noted that with smaller fractions of BOF slag, greater compressive strength values can be achieved. Used as an activator for blast furnace slag, activation reactions take place more slowly than when cement is used as an activator, and greater thickness and more binder are necessary to achieve the same strength characteristics. Nevertheless, binders based on slag develop good strength in the stabilized mixes and are suitable for use in road construction [19].

For a mixture of steel slag, fly ash, and phosphorus gypsum, the best features are achieved at the slag-to-fly ash ratio of 1:1, with the addition of phosphorous gypsum in the amount of 2.5% [20]. This mixture achieves the value of 7 days of compressive strength of 1.86 MPa, and 8.36 MPa for 28 days. Also, the modulus of elasticity is similar to that of the standard mixture, but with better resistance to water.

5.6 Slag utilization in asphalt mixes

One of the first applications of steel slag in asphalt mixtures took place in 1969 when a trial section was built in the Hamilton-Toronto area, in which steel slag was incorporated into binder and surface course asphaltic concrete mixes [23,24]. There have been no problems with the quality and durability of asphaltic concrete pavement in this instance, and any surface popouts observed were minor. In 1974, the Ministry of Transportation and Communications constructed 17 asphaltic concrete test sections in Ontario to evaluate methods for improving rigid pavement skid resistance with thin overlays for extremely high traffic volume. The resulting data indicated a fairly high polish stone value (PSV) compared to natural aggregates (for steel slag 55–59, for sandy dolomite 53, and for limestone 41–45), which led to the conclusion that steel slag asphaltic concrete exhibited good skid and wear resistance [23–25]. It was noted that the wet road accident rate did not significantly exceed the dry road accident rate for steel slag asphaltic concrete surfacing, while the reverse was generally true of other surfacing [24].

The application of steel slag in asphalt layers eventually became popular in the United Kingdom, Germany, Japan, and China. Common to all areas where steel slag was first used is their link to the developing steel industry.

During the first construction of asphalt surface layers with steel slag, cracks and surface popouts occurred, which was later linked to the instability of slag due to free oxides of calcium and magnesium. All these developments were due to a lack of careful material selection and improper construction [24,26,27].

Hydration of free oxides, which causes expansion, usually is prevented by having a sufficient weathering period before the use of slag in pavement layers or by using some other method of accelerated aging, as discussed in Chapter 4. However, it has been shown that aging is not the only criterion for the use of slag in asphalt layers if finer grain size fractions (13–13.2 mm) are used [23,24]. In the process of asphalt mix production, aggregate passes through the phases of washing, sieving, and drying, which causes the instantaneous hydration and expansion of slag.

Another condition that allows the application of slag without prior weathering is bituminous film coating of steel slag aggregate, which limits its potential for expansion [23]. For aggregates of larger granulation, especially those larger than 19 mm, a minimum weathering period of at least 30 days is required. As an additional measure to prevent deformation due to expansion, content of less than 3% by mass of nonslag constituents [26] and no more than 6% portion of free CaO [24,28] is recommended. The criterion of free CaO is significantly stricter if no weathered slag is used. In that case, the percentage of free CaO is limited to 4% for application in bituminous layers and 7% for application in unbounded base layers [19]. As an additional measure designed to prevent the possibility of pavement deformation during its service lifetime, Canadian regulations limit expansion to 1% [26].

Although the negative effects arising from the presence of free calcium oxide have been highlighted so far in this discussion, there also has been some positive impact on the application of slag in asphalt mixtures. The presence of free CaO increases the resistance of slag to stripping of bituminous binder and increases its adhesion between the aggregate and binder [26,29] which contributes to greater durability of the wearing courses. The increase in adhesion contributes to the very nature of steel slag as a by-product of steel production, which means that it cannot contain clay, which would reduce the adhesion [16].

Among the other advantages of steel slag for its use in asphalt mixtures, its physical characteristics should be mentioned as well. Sharp edges, proper grain shape, and rough surface texture also improve the adhesion of binder and aggregate and increase the coefficient of friction in bituminous surface courses over any natural aggregate. These physical characteristics, with proper particle size distribution, increase the shear strength and resistance to rutting [16]. This is the reason that steel slag asphalt mixtures are used most often in places where high abrasion resistance and good adhesion are required, such as industrial facilities, parking areas exposed to heavy traffic, and intersections.

High specific gravity and density of steel slag are among the highlighted physical properties, which are responsible for the increase in specific gravity and density of bituminous mixtures compared with natural aggregates [23,25,26]. The high specific gravity is an advantage in road maintenance during the winter. Also, a high stability of mixtures containing steel slag as an aggregate (which is up to 1.5–3 times higher than the standard mixture [23,26,29]) and excellent heat retention (except for a reduction in energy consumption during winter patching) contribute to patch durability. During construction, these mixtures showed great workability and heat retention and very good compaction properties [16,23,25,26].

Confirmation of the suitability of steel slag for use in asphalt mixtures was supplied by research conducted at McMaster University, which found a 20%–80% greater modulus of elasticity of an asphalt mixture containing steel slag than that of a standard aggregate at a temperature of 29°C [23,24]. Tests have also shown lower values of the

Poisson coefficient than a standard mixture, which indicates greater asphalt mixture stability and resistance to plastic deformation [24]. Due to the given data and the pavement structure stress state, the possibility of designing pavement structures with reduced asphalt concrete thicknesses is indicated [23]. Hereby, compensation for one of the drawbacks of this mixture (namely, 15%–25% higher volume weight, which increases the cost of transportation of fresh asphalt mixtures) could be established. Of these drawbacks, an increased need for binders due to steel slag aggregate sharp edges and rough surfaces can be highlighted.

Steel slag as an aggregate can be used in both hot and cold asphalt mixtures. As the only limitation for its use, the amount of steel slag aggregate consisted in the mixture is stated. Hot mix asphalt containing 100% steel slag is susceptible to high void space and bulking problems due to the angular shape of steel slag aggregate. Mixes with high void space (100% steel slag aggregate mixes) are susceptible to overasphalting during production and subsequent flushing due to in-service traffic compaction [26].

Asphalt mixes containing steel slag required higher optimum proportion of bitumen compared to the asphalt with natural aggregate. The amount of bitumen in mixes with the natural aggregate ranges from 4.5%–5.0%, while in mixtures with slag added, it ranges from 4.7%–5.8% [30,31]. This is a result of high steel slag porosity, which lead to excessive asphalt absorption, which in turn leads to a high cost of asphalt production [32]. Another reason for increased binder content in slag asphalt can be a loose stack of impurities on the surface of steel slag aggregate.

For this reason, it is recommended the use of slag only as a replacement for finer or coarser aggregate fractions with applications in wearing asphalt layers. Table 5.1 shows the values of the optimum binder content for different types of aggregates.

Asphalt concrete mixtures have the highest abrasion resistance if they contain 30% of slag aggregate, while 75% replacement of limestone by slag aggregates results in significant improvement in the mechanical properties of asphalt concrete [30]. Among all properties, Marshall stability is especially highlighted, which in this study ranged from 19.54–20.19 kN for asphalt concrete, while for the asphalt with natural aggregate ranged from 16.50–17.46 kN. That results in greater resistance to permanent deformation of such surface layers and greater stiffness. Asphalt with slag aggregate shows 1.9–2.4 times greater values of indirect tensile stiffness modulus for all test temperatures, especially 20°C.

Table 5.1 Optimal binder content in asphalt mix depending on aggregate type [24,28,29]

Type of mixture	% Weight of total mix
Asphalt concrete with natural aggregate	4.5–4.6
Asphalt concrete with slag	5.5–6
Porous asphalt with BOF slag	5.5–5.76
Slag with sandy limestone	4.7–4.8

In order to further improve slag utilization in asphalt concrete, modified steel slag aggregate preparation can be adopted. As stated earlier in this chapter, steel slag aggregate for use in asphalt mixture production can have impurities on its surface that affect final product quality. The modified method of slag aggregate preparation consisted of vibrating sieve, drum sieve, and high-pressure water jet treatment. Using this method, higher fine aggregate angularity and sand equivalent aggregate can be obtained [32]. Utilizing this modified aggregate, improved mechanical properties and deformation resistance, greater cohesive strength and moisture resistance, lower rutting depth, and higher dynamic stability of asphalt mixture can be achieved.

Testing of electrical conductivity also confirms slag as a better aggregate comparing to natural materials. Electrical conductivity can be useful in special techniques of winter maintenance (e.g., airport runway maintenance). When measuring skid resistance with pendulum skid resistance test, higher skid number (SN) values for asphalt concrete mixtures containing 30% slag aggregate (SN ≈ 100) than those of ordinary asphalt concrete (SN ≈ 85) [31].

Unlike applications in asphalt concrete, steel slag, particularly that produced by the base oxygen procedure (i.e., BOF slag) by up to 100% of aggregate in porous asphalt, can be replaced. Compared with porous asphalt containing an aggregate of crushed river stone, porous asphalt with BOF slag has several advantages. These asphalts are less susceptible to abrasion and have better properties in terms of traffic noise emissions [29]. Better absorption due to high porosity and excellent surface water drainage, which contribute to higher traffic safety, are also shown in this investigation. BOF slag has a great particle shape and rough surface texture that exhibits a good interlocking mechanism and internal friction in porous asphalt. It all contributes to higher Marshall stability. Increasing the share of BOF slag in porous asphalt increases skid resistance and the stripping resistance due to the presence of lime.

Application of slag in asphalt mixture for rigid pavement rehabilitation was constructed in the province of Hubei, China [28]. Rigid pavement rehabilitation included a build of a new split mastic asphalt (SMA) layer for the construction of a 400-m-long road section. Expansion of the embedded layer was very slow; after 7 days, it reached only 1%, while the adhesion bond between the aggregate and binder were extremely good. This increases the wear resistance of the pavement so that it can endure higher traffic loads and higher temperature influences. Better resistance to permanent deformation at high temperatures and better crack resistance at low temperatures were also shown.

In order to improve the road surface properties, asphalt layer containing BOF slag was built on a 2-km-long highway section. After two years of service, it showed excellent behaviour due to the emergence of cracks and rut (Table 5.2) [33].

In addition to all positive impact of steel slag application in asphalt mixtures, the financial feasibility of using this material should be stated [30]. However, the use of BOF slag in asphalt mixtures has disadvantages as well. For example, the curing period of the asphalt surface layer before opening to traffic takes longer. In fact, with larger amounts of BOF slag, the heat inside the pavement takes longer to be released, and a longer stabilization time was needed to reach thermal equilibrium [34]. Since BOF slag is a porous material that can store heat, the accumulated heat was then

Table 5.2 **Properties of asphalt surface layer containing BOF slag [33]**

Property	Exploitation period			
	6 Months	12 Months	18 Months	24 Months
Bulk density of core (g/cm^3)	2.511	2.520	2.524	2.525
Abrasion and friction coefficient (BPN)	62	60	56	55
Surface texture depth (mm)	1.2	0.9	0.8	0.8

convected upward to replenish energy in the upper layer of the pavement. As a result, the cooling rate decreased in the pavement containing BOF slag.

5.7 Slag utilization in concrete pavement

Concrete that is built into rigid pavement does not differ significantly from concrete used in building construction. The properties of ordinary concrete containing slag were discussed earlier in this chapter; in this section, the essential characteristics of concrete for use in pavement structures are presented.

In the construction of industrial concrete pavement, except for high strength and high resistance to wear, chemical influence and durability are also preferred concrete properties. These materials are often very expensive, and multiple researchers have been trying to find new ways to improve the mechanical properties of concrete for use in industrial pavement. Therefore, concrete with different aggregate types have been compared: steel slag, limestone, and river sand, and with a small cement portion (270 kg/m^3), a low water-cement ratio (0.52), and a maximum aggregate grain size of 31.5 mm [35]. Mixtures with slag aggregate have showed better mechanical properties then concrete with natural aggregate. Coarse slag aggregates show a nearly 50% improvement in wear resistance, which is especially important in concrete for industrial pavement. Adding steel slag aggregate causes a more than 40% increase in compressive and flexural strength over concrete with limestone aggregate, and these concrete mixtures can meet the requirements for industrial pavement. This provides an opportunity to preserve natural resources and achieve environment conservation by using this waste material in concrete production.

The state-of-the-art example of steel slag usage as an aggregate in concrete pavement is a situation in Belgium where slag from stainless steel production is used as an aggregate in roller-compacted concrete pavement construction of a rural road with very light traffic [18]. Steel slag aggregate of 0/30-mm granulation was used, to which sand of 0/4 mm granulation was added to improve the workability and disposition of concrete. Before a test section was built, laboratory studies were conducted in which concrete with aggregate consisting of 85% slag, 15% sand, and 16% cement by aggregate weight was used. The study included 7-, 14-, and 28-day compressive strength measurement. After 7 curing days, the mean compressive strength of three samples was 26.7 MPa; after 14 days, 31.3 MPa; and after 28 days, 37.1 MPa. These findings encouraged continued research. Since aggregate segregation was observed on some samples, the composition of concrete that is embedded into the test section was partially altered by increasing the amount of sand to 20% and reducing the quantity of slag to 80%. Laboratory tests of embedded concrete showed mean compressive strength of 29 MPa at 7 days, mean compressive strength of 31.6 MPa at 32 days, and mean compressive strength of 36.2 MPa at 90 days, which was rated as very good.

5.8 Slag utilization in hydrotechnical structures

The high density of steel slag characterized it as a good structural material for hydrotechnical structures. In Germany and the Netherlands, about 400,000 tons of this material per year is used for slope protection. Specifically, a coarse aggregate is used to protect against high water [1]. The size of the aggregates ranges from 100–250 mm. Used for this purpose, the aggregate should have acceptable volume stability and be wear resistant. Steel slag with free lime content under 5%, or such material that has been weathered for at least three months, is considered to have acceptable volume stability [1]. In addition to slope lining, this material is used for water flow regulation and eroded riverbed reclamation to direct water into a navigable riverbed during the dry season [36].

5.9 Slag utilization for acid mine drainage treatment

Acid mine drainage or *acid and metalliferous drainage (AMD)* refers to the outflow of acidic water from metal mines or coal mines. AMD forms when sulfide minerals (particularly pyrite and marcasite) have been exposed to oxidizing conditions during mining and other excavation activities such as highway construction [37]. In the presence of oxygen and water, sulfide minerals oxidize to form sulfate-rich and often metal-laden drainage. Untreated AMD flowing into streams can degrade habitat and water quality, often producing an environment devoid of most aquatic life and unfit for use. In addition, AMD can be toxic to vegetation and can reduce the potability of water supplies [37]. Different active and passive water treatment techniques have been developed during the last several decades to control AMD. Passive treatment of AMD has largely featured the dissolution of limestone to increase pH and alkalinity and precipitate metals. However, limestone is susceptible to armouring due to metal precipitation and has limited effectiveness in highly acidic or high flow discharges. Unlike lime, steel slags do not absorb CO_2 from the air and convert back to relatively insoluble calcite [38]. Some constituents of the slag (e.g., $SiO3^{2-}$ and CaO) can neutralize acidity, of which silicic acid is less harmful to the environment than carbonic acid. Consequently, steel slag can be a suitable alternative to limestone for the neutralization of AMD, and results shown in [37–42] indicate that steel slag can provide highly concentrated alkaline recharge to acid mine waste.

References

[1] H. Motz, Production and use of air-cooled blastfurnace and steel slags, in: J. Geiseler, M. Dean (Eds.), 3rd European Slag Conference: Proceedings Manufacturing and Processing of Iron and Steel Slags, EUROSLAG Publication, Keyworth, UK, 2002, pp. 7–20.

[2] EUROSLAG, Statistics. Accessed February 21, 2014, from http://www.euroslag.org/products/statistics.

[3] Wordsteel Association, World steel in figures 2014, Accessed February 21, 2015, from https://www.worldsteel.org/dms/internetDocumentList/bookshop/World-Steel-in-Figures-2014/document/World%20Steel%20in%20Figures%202014%20Final.pdf

[4] I. Arribas, et al., Application of steel slag concrete in the foundation slab and basement wall of the TECNALIA Kubik building, in: H. Motz, P. Verhaag (Eds.), 6th European Slag Conference: Ferrous Slag—Resource Development for an Environmentally Sustainable World, EUROSLAG Publication, Madrid, Spain, 2010, pp. 251–64.

[5] G. Bernardo, et al., The use of oil well-derived drilling waste and electric arc furnace slag as alternative raw materials in clinker production, Res. Conserv. Recycl. 52 (2007) 95–102.

[6] P.E. Tsakiridis, et al., Utilization of steel slag for Portland cement clinker production, J. Haz. Mat. 152 (2008) 805–11.

[7] R.I. Iacobescu, et al., Ladle metallurgy stainless steel slag as a raw material in ordinary Portland cement production: A possibility for industrial symbiosis, J. Cleaner Prod. 112 (2) (2016) 872–881.

[8] H. Qasrawi, F. Shalabi, I. Asi, Use of low CaO unprocessed steel slag in concrete as fine aggregate, Const. Bldg. Mat. 23 (2) (2009) 1118–1125.

[9] I. Arribas, et al., Electric arc furnace slag and its use in hydraulic concrete, Const. Bldg. Mat. 90 (2015) 68–79.

[10] M. Maslehuddin, et al., Comparison of properties of steel slag and crushed limestone aggregate concretes, Const. Bldg. Mat. 17 (2003) 105–112.

[11] K. Onoue, et al., Fatigue characteristics of steel-making slag concrete under compression in submerged condition, Const. Bldg. Mat. 70 (2014) 231–242.

[12] B. Pang, Z. Zhou, H. Xu, Utilization of carbonated and granulated steel slag aggregate in concrete, Const. Bldg. Mat. 84 (2015) 454–467.

[13] R. Baciocchi, et al., Effctes of thin-film accelerated carbonation on steel slag leaching, J. Haz. Mat. 286 (2015) 369–378.

[14] N. Faraone, et al., Steelmaking slag as aggregate for mortars: Effects of particle dimension on compression strength, Chemosphere 77 (2009) 1152–1156.

[15] H. Motz, J. Geiseler, Products of steel slags an opportunity to save natural resources, Waste Manage. 21 (2001) 285–293.

[16] National Slag Association, Steel slag: A premier construction aggregate, Accessed February 21, 2014, from http://www.acobrasil.org.br/siderurgiaemfoco/CCABrasil/NSA%20Steel%20Furnace%20Brochure.pdf.

[17] V.A. Mymrin, H.A. Ponte, C.I. Yamamoto, Synthesis of new colloidal formations during the strengthening of different activated hydrated metallurgical slags, Coll. Surf. A: Physicochem Eng Asp. 220 (2003) 211–221.

[18] L.P. De Bock, H. Van de Bergh, Stainless steel slags in hydraulic bound mixtures for road construction, two case studies in Belgium, Accessed February 24, 2014, from www.congress.cimne.upc.es/rilem04/admin/Files/FilePaper/p295.pdf.
[19] M. Mäkikyrö, Converting raw materials into the products—Road base material stabilized with slag-based binders, Doctoral Dissertation Department of Process and Environmental Engineering, University of Oulu, Oulu, Finland, 2004.
[20] W. Shen, et al., Investigation on the application of steel slag–fly ash–phosphogypsum solidified material as road base material, J. Haz. Mat. 164 (2009) 99–104.
[21] M. Luckman, V. Satish, D. Venkateswaran, Cementitious and pozzolanic behavior of electric arc furnace steel slags, Cement Concr. Res. 39 (2009) 102–109.
[22] P.Y. Mahieux, J.E. Aubert, G. Escadeillas, Utilization of weathered basic oxygen furnace slag in the production of hydraulic road binders, Const. Bldg. Mat. 23 (2009) 742–747.
[23] J. Emery, Slag utilization in pavement construction, in: W.W. Hotaling(Ed.), Extending aggregate resources, ASTM, Baltimore, USA, 1982, pp. 95–119.
[24] J. Emery, Steel slag utilization in asphalt mixtures, Accessed February 10, 2015, from http://www.nationalslag.org/sites/nationalslag/files/documents/nsa_186-1_steel_slag_utilization_in_asphalt_mixes.pdf.
[25] N.A. Ali, et al., The use of steel slag in asphaltic concrete, in: R.C. Meininger (Ed.), Effects of Aggregates and Mineral Fillers on Asphalt Mixture Performance, American Society for Testing and Materials, Philadelphia, USA, 1992, pp. 3–19.
[26] User guidelines for byproducts and secondary use materials in pavement construction. Steel slag. Publication Number: FHWA-RD-97–148. Accessed February 16, 2015, from https://www.fhwa.dot.gov/publications/research/infrastructure/structures/97148/intro.cfm
[27] N.O.D. Silveira, et al., ACERITA—Steel slag with reduced expansion potential, in: H. Motz, S. Haimi, M. Mäkikyrö (Eds.), 4th European Slag Conference: Slags—Providing Solutions for Global Construction and Other Markets, EUROSLAG Publication, Oulu, Finland, 2005, pp. 145–157.
[28] S. Wu, et al., Utilization of steel slag as aggregates for stone mastic asphalt (SMA) mixtures, Bldg. Environ. 42 (2007) 2580–2585.
[29] D.-H. Shen, C.-M. Wu, J.-C. Du, Laboratory investigation of basic oxygen furnace slag for substitution of aggregate in porous asphalt mixture, Construct. Bldg. Mat. 23 (2009) 453–461.
[30] P. Ahmedzade, B. Sengoz, Evaluation of steel slag coarse aggregate in hot mix asphalt concrete, J. Haz. Mat. 165 (2009) 300–305.
[31] F. Kehagia, Skid resistance performance of asphalt wearing courses with electric arc furnace slag aggregates, Waste Manage. Res. 27 (2008) 288–294.
[32] Z. Chen, et al., Utilization of gneiss coarse aggregate and steel slag fine aggregate in asphalt mixture, Const. Bldg, Mat. 93 (2015) 911–918.
[33] Y. Xue, et al., Experimental investigation of basic oxygen furnace slag used as aggregate in asphalt mixture, J. Haz. Mat. B138 (2006) 261–268.
[34] L.-S. Huang, et al., In-situ temperature effects in basic oxygen furnace slag asphalt concrete pavement, Intl. J. Pavement Res. Tech. 6 (4) (2013) 386–394.
[35] I. Papayianni, E. Anastasiou, M. Papachristoforou, High performance concrete for pavements with steel slag aggregate, in: J. Zachar, P. Claisse, T.R. Naik, E. Ganjian (Eds.),Second International Conference on Sustainable Construction Materials and Technologies, Università Politecnica delle Marche, Ancona, Italy, 2010.
[36] M. Joost, The utilisation of steel slag in waterway construction, in: J. Geiseler, M. Dean (Eds.), 3rd European Slag Conference: Proceedings Manufacturing and Processing of Iron and Steel Slags, EUROSLAG Publication, Keyworth, UK, 2002, pp. 17–107.

[37] J.S. Simmons, P. Ziemkiewicz, An alternative alkaline addition for direct treatment of acid mine drainage, in: Proceedings of the Sudbury 2003 Mining and the Environment Conference, May 2003, Sudbury, Canada, Laurentian Univ., Sudbury, Canada, Sudbury, Canada, 2003.

[38] P. Ziemkiewicz, Steel slag: Applications for AMD control, in: L.E. Erickson, M.M. Rankin (Eds.), Proceedings of the 1998 Conference on Hazardous Waste Research, Kansas State University, Manhattan, USA, 1998, pp. 44–62.

[39] C. Leite, L.P. Cardoso, J.M.W. Mello, Use of steel slag to neutralize acid mine drainage (AMD) in sulfidic material from a uranium mine, Revista Bras. de Ciência do Solo 37 (3) (2013) 804–811.

[40] B. Mack, B. Gutta, An analysis of steel slag and its use in acid mine drainage (AMD treatment, ASMR, in: R.I. Barnhisel (Ed.), 2009 National Meeting of the American Society of Mining and Reclamation: Revitalizing the Environment, Proven Solutions and Innovative Approaches Lexington, KY, USA, 2009, pp. 42–722.

[41] D. Kirby, Effective treatment options for acid mine drainage in the coal region of West Virginia, Master Thesis, Graduate College of Marshall University, Marshall University, 2014.

[42] T.I. Name, Remediation of acid mine drainage using metallurgical slags, Master Thesis, Faculty of Engineering and the Built Environment, University of the Witwatersrand, Johannesburg, 2013.

The Croatian experience of steel slag application in civil engineering

6

Two slag landfills are located near the cities of Sisak and Split in Croatia. Air-cooled slag from steel production (originally electric arc) is deposited in both croatian landfils. The amount of slag deposited near the Sisak plant is estimated to be 1.5 million tons spread over 25 ha, while the amount of slag deposited in the Split Ironworks landfill is estimated to be 30,000 tons. Those landfills are shown in Figures 6.1 and 6.2. Currently, this type of slag has limited application, primarily in the construction of field paths and, for soil improvement.

Foreign research cites many applications for blast furnace slag, which have been incorporated into regulations and tested by long experience with cement production. Because of its high density and hardness, steel slag is used considerably less; however, the high cost of slag disposal dictates the need to find new areas for its utilisation. Next, the results of investigations of Croatian steel slag uses in various fields of civil engineering are described.

6.1 Slag as a structural concrete aggregate

The possibility of utilising steel slag as an aggregate in concrete was investigated in the project E!4166—EUREKABUILD FIRECON, financed by the Croatian Ministry of Education, Science, and Sport. The project centred on the properties of hardened concrete made with steel slag aggregates from the landfills near Sisak and Split, under both normal exposure and fire conditions.

Due to its condition as disposed waste, slag available in Croatia typically cannot be used as an aggregate in concrete, and preparatory actions had to be undertaken before testing. These actions included manual selection of waste and coarse units of slag, manual sieving of materials and their separation into fractions (0–4 mm, 4–8 mm, 8–16 mm, and >16 mm), watering of aggregate fractions designated for analysis (4–8 mm and 8–16 mm), and manual selection using a magnet to collect aggregate particles containing metals (Figures 6.3–6.8). Slag separated into fractions is shown in Figures 6.9 and 6.10.

Properties important for using slag as an aggregate in concrete (Table 6.1) were determined on coarse fractions (4–8 mm and 8–16 mm) in accordance with the standard EN 12620/AC:2006 [1], with volume stability testing as the only exception. In accordance with EN 1744-1:2004 [2], testing of aggregate volume stability was performed on mortar specimens prepared with an aggregate with a grain size of 0–20 mm and cement, and based on the results, slag was classified

Characteristics and Uses of Steel Slag in Building Construction. http://dx.doi.org/10.1016/B978-0-08-100368-8.00006-3

Figure 6.1 Landfill near Sisak

Figure 6.2 The Split ironworks landfill

Figure 6.3 Slag in its natural state

Figure 6.4 Coarse units of slag that has been manually selected

Figure 6.5 Waste that has been manually selected out of slag

Figure 6.6 Sieving of slag

Figure 6.7 Using a magnet to select aggregate particles containing metals by hand

Figure 6.8 Washing slag

Figure 6.9 Slag as an aggregate, mix of fractions 4-8 and 8-16 mm

Figure 6.10 Slag as an aggregate, fraction 0–4 mm

Table 6.1 Testing carried out in accordance with EN 12620/AC:2006 [4]

Property	Reference standard
Particle size distribution	EN 933-1:2003
Fines content	EN 933-1:2003
Particle shape—shape index	EN 933-4:2008
Apparent particle density	EN 1097-6:2002/AC:2004/A1:2005
Particle density on an oven-dried basis*	EN 1097-6:2002/AC:2004/A1:2005
Particle density on a saturated and surface-dried* basis	EN 1097-6:2002/AC:2004/A1:2005
Water absorption	EN 1097-6:2002/AC:2004/A1:2005
Petrographic description	EN 932-3:2003
Volume stability—drying shrinkage	EN 1367-4:2008
Chloride content**	EN 1744-1:2004
Total sulphur content**	EN 1744-1:2004
Acid-soluble sulphate content**	EN 1744-1:2004
Dicalcium silicate disintegration	EN 1744-1:2004
Iron disintegration	EN 1744-1:2004
Resistance to fragmentation—Los Angeles test	EN 1097-2:2004
Freeze-thawing resistance	EN 1367-2:2004

*By pycnometer
**Not necessary if complete chemical analysis is obtained

in accordance with EN 12620/AC and then compared to the Croatian regulatory requirements [3]. The test results for each type of slag are given in Tables 6.2 and 6.3. The amounts of free CaO and MgO in slag are listed in Table 6.4.

Considering the results shown in Tables 6.2 and 6.3, it can be concluded that both types of slag satisfy the Croatian technical requirements for reinforced concrete structures [3] with one exception: the volume stability of the specimen made with Slag 2.

Table 6.2 **Test results for slag 1 (Sisak) [4]**

Analysed property	Fraction	Class in accordance with EN 12620/property value/ description	Requirement in accordance with croatian technical regulation for concrete structures
Particle size distribution	4–8 mm	Class $G_C85/20$	Requested class: $G_C85/20$
	8–16 mm	Class $G_C90/15$	Requested class: $G_C90/15$
Fines content	4–8 mm	Class $f_{1,5}$	Highest class permitted: $f_{1.5}$
	8–16 mm	Class $f_{1,5}$	Highest class permitted: $f_{1.5}$
Particle shape—shape index	4–8 mm	Class SI_{15}	Highest class permitted: SI_{20}
	8–16 mm	Class SI_{15}	Highest class permitted: SI_{20}
Apparent particle density	4–16 mm	3.49 Mg/m^3	Requirements of the client and customer must be satisfied.
Particle density on an oven-dried basis		3.09 Mg/m^3	
Particle density on a saturated and surface-dried basis		3.21 Mg/m^3	
Water absorption		3.7%	
Grain shape	4–16 mm	Isometric to elongated	Requirements of the client and customer must be satisfied.
Appearance of the surface		Rough, spongelike	
Roundness of the grain		Square	
Fragility		No	
Volume stability—drying shrinkage	0–20 mm	0.026%	Permitted: 0.075%
Chloride content	4–16 mm	0%	Permitted: 0.06% for reinforced concrete
Acid-soluble sulphate content		Class $AS_{1,0}$	Highest class permitted: $AS_{1.0}$
Total sulphur content		0.3%	Permitted: 2%
Dicalcium silicate disintegration	4–16 mm	Not present	Not permitted
Iron disintegration		Not present	
Resistance to fragmentation—Los Angeles test	4–16 mm	Class LA_{25}	Highest class permitted: LA_{35} for concrete of common allocation; LA_{30} for concrete in XF1 to XF4 class of exposure (according to EN 206-1)
Freeze-thawing resistance	4–16 mm	Class MS_{18}	Highest class permitted: MS_{25} for concrete in XF1 to XF3 class of exposure; MS_{18} for concrete in XF2 to XF4 class of exposure (according to EN 206-1)

Table 6.3 **Test results for slag 2 (Split) [4]**

Analysed property	Fraction	Class in accordance to EN 12620/property value/ description	Requirement in accordance with TPBK
Particle size distribution	4–8 mm	Class $G_C85/20$	Requested class: $G_C85/20$
	8–16 mm	Class $G_C90/15$	Requested class: $G_C90/15$
Fines content	4–8 mm	Class $f_{1.5}$	Highest class permitted: $f_{1.5}$
	8–16 mm	Class $f_{1.5}$	Highest class permitted: $f_{1.5}$
Particle shape – shape index	4–8 mm	Class SI_{15}	Highest class permitted: SI_{20}
	8–16 mm	Class SI_{15}	Highest class permitted: SI_{20}
Apparent particle density	4–16 mm	3.41 Mg/m^3	Requirements of the client and customer must be satisfied.
Particle density on an oven-dried basis		2.93 Mg/m^3	
Particle density on a saturated and surface-dried basis		3.07 Mg/m^3	
Water absorption		4.9%	
Grain shape	4–16 mm	Isometric to elongated	Requirements of the client and customer must be satisfied
Appearance of the surface		Rough, spongelike	
Roundness of the grain		Square to slightly rounded	
Fragility		no	
Volume stability—drying shrinkage	0–20 mm	Specimen did not harden	Permitted: 0.075%
Chloride content	4–16 mm	0%	Permitted: 0.06% for reinforced concrete
Acid-soluble sulphate content		Class $AS_{1.0}$	Highest class permitted: $AS_{1.0}$
Total sulphur content		0.48%	Permitted: 2%
Dicalcium silicate disintegration	4–16 mm	Not present	Not permitted
Iron disintegration		Not present	
Resistance to fragmentation—Los Angeles test	4–16 mm	Class LA_{25}	Highest class permitted: LA_{35} for concrete of common allocation; LA_{30} for concrete in XF1 to XF4 class of exposure (according to EN 206-1)
Freeze-thawing resistance	4–16 mm	Class MS_{18}	Highest class permitted: MS_{25} for concrete in XF1 to XF3 class of exposure; MS_{18} for concrete in XF2 to XF4 class of exposure (according to EN 206-1)

Table 6.4 **Content of free CaO and MgO in slag [4]**

Applied method	Analysed property	Property value (% of mass)	
		Slag 1	Slag 2
Combination of complexometric method (described in EN 1744-1) and X-ray diffraction	Content of free CaO	0.28	0.22
X-ray diffraction	Content of free MgO	4.82	4.21

This specimen would not harden, so it was not possible to evaluate the volume stability. The inability of this specimen to harden was assumed to have been due to partial dicalcium silicate (C_2S) disintegration. C_2S disintegration could not be observed by testing, according to EN 1744-1:2004 [2], because of the limitation of the method used. A more detailed or sophisticated analysis of C_2S in this slag would probably have shown a higher proportion of γ C_2S (the unstable form of C_2S) than β C_2S (stable form C_2S).

Given that γ C_2S has low hydraulic properties [5], its presence in fine particles of slag could explain the inability of the concrete to harden. However, washing the slag aggregate and using only coarse slag fractions (4–8 mm and 8–16 mm) are assumed to prevent this problem. Therefore, only coarse slag fractions (4–8 mm, 8–16 mm) are used in concrete.

In further research, three groups of concrete specimens were created, each with a different concrete mixture (M, M1, and M2). The three mixtures were prepared with the same cement type (CEM I 52,5N) and content (400 kg/m^3) and the same water-to-cement ratio (w/c = 0.43), and superplasticiser (0.8% of the cement mass) was added. The reference mixture (M) was completely prepared with dolomite (fractions 0–4 mm, 4–8 mm, and 8–16 mm). The other two mixtures, M1 and M2, were made with slag available within Croatia, which was used as a coarse aggregate (4–8 mm and 8–16 mm), and with dolomite as a fine aggregate (0–4 mm). The cumulative sieving curve (C) of aggregates, obtained by mixing different aggregate fractions in precisely prescribed proportions, was the same for all three mixtures (Figure 6.11). A suitable overall grading of the aggregate (C) was determined by values lying between those two recommended curves (i.e., the Fuller and EMPA curves), which provide good aggregate packing and concrete properties [6].

The chemical characteristics of cement and aggregates are presented in Table 6.5. The properties of dolomite used for concrete preparation (the reference mixture) are given in Table 6.6.

Aggregates used for preparing concrete were first saturated and then surface dried. The process was performed by placing aggregates in a water tank for 24 h and then removing them and wiping excess water from the surface.

Details of the mixture are given in Table 6.7, and the properties of prepared concrete mixtures in their fresh state are given in Table 6.8. All testing of fresh concrete was performed according to the relevant European standards; density was measured

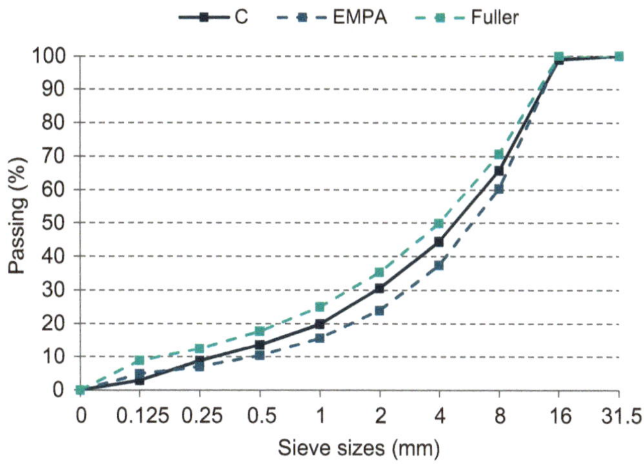

Figure 6.11 Cumulative sieving curve [4]

Table 6.5 Chemical properties of materials (% of mass) [4]

Compound/material	Cement	Aggregates		
		Dolomite	Slag 1	Slag 2
SiO_2	19.74	0.49	17.08	14.24
CaO	63.35	31.78	24.98	31.52
Al_2O_3	5.33	0.10	5.40	7.60
Fe_2O_3	2.25	0.10	25.45	25.74
MgO	2.72	20.85	10.58	7.42
MnO	0.18	–	8.91	3.80
Na_2O	0.3	0.01	0.12	0.13
K_2O	0.84	0.01	0.13	0.08
SO_3^{2-}	–	–	0.25	0.44
S^{2-}	–	–	0.05	0.04
SO_3	3.3	–	–	–
TiO_2	–	–	–	–
Loss on ignition (LOI)	1.34	46.64	4.99	4.71

according to EN 12350-6:2009 [7], air content according to EN 12350-7:2000 [8], and consistency according to EN 12350-2:2000 [9]. For estimating the development of flexural and compressive strength over time, prisms measuring 10 × 10 × 40 cm were tested. Volume changes of concrete and the static modulus of elasticity were determined on prisms with the same measurements. For defining a concrete class, cubes of 15 cm in the edge length were tested at 28 days.

All specimens were demoulded 24 h after casting, placed in a water tank for 7 days (Figure 6.12), and placed in a chamber with a temperature of about 20 ± 2 °C and a humidity level of 95 ± 5% for another 27 days. After 28 days, some specimens were

Table 6.6 **Test results for dolomite [4]**

Analysed property	Fraction	Property value
Apparent particle density	0–4 mm	2.79 Mg/m^3
	4–16 mm	2.79 Mg/m^3
Particle density on an oven-dried basis	0–4 mm	2.73 Mg/m^3
	4–16 mm	2.72 Mg/m^3
Particle density on a saturated and surface-dried basis	0–4 mm	2.75 Mg/m^3
	4–16 mm	2.75 Mg/m^3
Water absorption	0–4 mm	0.9%
	4–16 mm	0.8%

tested, and others were left in the laboratory room for a total of 90 days. The specimens intended to be used for the measurement of volume changes were left in the chamber for 91 days. For corrosion susceptibility testing, a concrete specimen in the fresh state was placed in a closed mould and kept there for 28 days with no air or water exposure to avoid carbonation (Figure 6.13).

Flexural strength testing was performed according to EN 12390-5:2009 [10], and static modulus of elasticity testing was done according to HRN U.M1.025:1982 [11]. To reduce the number of specimens, the rate of compressive strength development was tested on the prism halves that remained after flexural strength testing. The compressive strength of the 28-day-old concrete was tested according to EN 12390-3:2002 [12]. All the testings are shown in Figures 6.14–6.17.

The possible corrosive effects of such concrete on reinforcing elements were estimated by testing the pH value of the solution prepared with powder and distilled water, according to [13]. The pore solution used in experiments was produced by grinding and collecting concrete powder from hardened concrete specimens and mixing it with distilled water in a 1:10 ratio. After 24 h, the solution was passed through filter paper and the pH value of the filtrate was measured with a pH meter (Figure 6.18).

To estimate the possibly high rate of slag expansion, volume changes of slag concrete were compared with volume changes of reference concrete made with dolomite, the commonly used natural aggregate. Concrete volume changes were measured according to HRN U.M1.029:1983 [14]. The specimens that were measured for volume changes are shown in Figure 6.19. Volume changes were examined on three prisms for each mixture and two measuring lines on each prism.

The compressive strength of the 28-day-old concrete cubes is shown in Figure 6.20. The flexural and compressive strengths of the observed specimens, which were 2, 28, 56, and 90 days old, are given in Figures 6.21 and 6.22. The values achieved for the modulus of elasticity are shown in Figure 6.23. Each result is the average of testing three specimens for each property.

Because the same amount of cement was used in all three mixtures, along with the same w/c ratio, the same amount of superplasticiser and the same cumulative sieving curve of aggregates, the results demonstrate the crucial impact of an aggregate type on concrete mechanical properties. At 28 days old, the concrete with Slag 1 (M1) had

Table 6.7 Mixture details [4]

Mixture	w/c	Superplasticiser (kg)	Cement (kg)	Aggregate (kg)						
				Dolomite, fraction 0–4 mm	Dolomite, fraction 4–8 mm	Dolomite, fraction 8–16 mm	Slag 1, fraction 4–8 mm	Slag 1, fraction 8–16 mm	Slag 2, fraction 4–8 mm	Slag 2, fraction 8–16 mm
M	0.43	3.20	400	807	367	661	–	–	–	–
M1	0.43	3.20	400	864	–	–	422	723	–	–
M2	0.43	3.20	400	841	–	–	–	–	411	704

Table 6.8 Properties of fresh concrete [4]

Mixture	Density of fresh concrete (kg/m^3)	Air content (%)	Consistency by slump test (cm)
M	2530	0.9	20
M1	2643	2.0	20
M2	2574	2.3	20

Figure 6.12 Curing of samples

Figure 6.13 Sample of fresh concrete mixture for testing the pH value of concrete eluate

86% of the compressive strength of the reference mixture (M), while the concrete with Slag 2 (M2) had 69% of the compressive strength of the M mixture (Figure 6.20). Figure 6.21 illustrates that both types of slag yield concrete with lower flexural strength in comparison with the reference mixture. The reduction of flexural strength of all 56-day-old mixtures, shown in Figure 6.21, has no valid explanation. The lower values

Figure 6.14 Flexural strength testing

Figure 6.15 Compressive strength testing on prism halves

Figure 6.16 Compressive strength testing on concrete cubes with 15-cm edge lengths

Figure 6.17 Modulus of elasticity testing

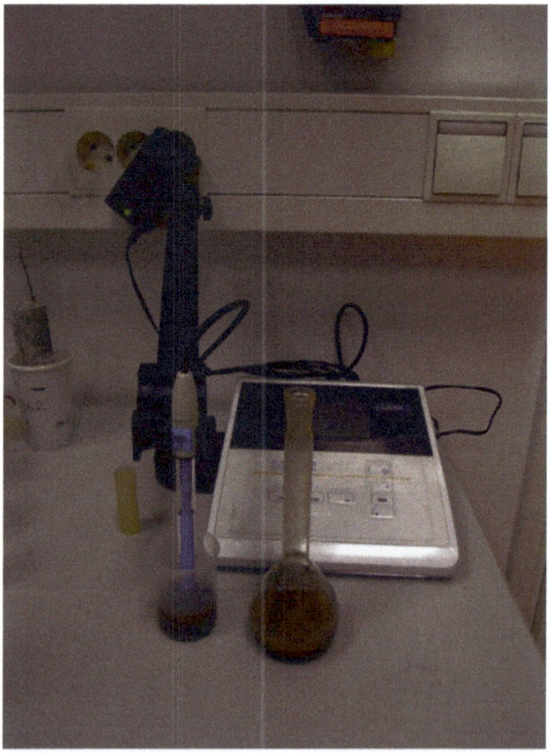

Figure 6.18 Measuring of the pH value of concrete eluate

Figure 6.19 Volume change testing

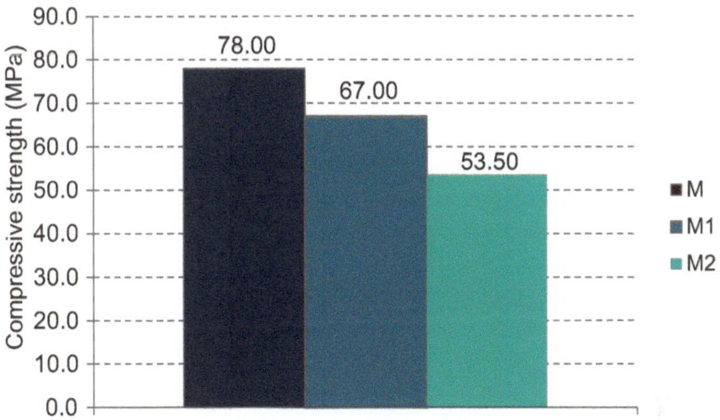

Figure 6.20 Compressive strengths of concrete cubes at the age of 28 days [4]

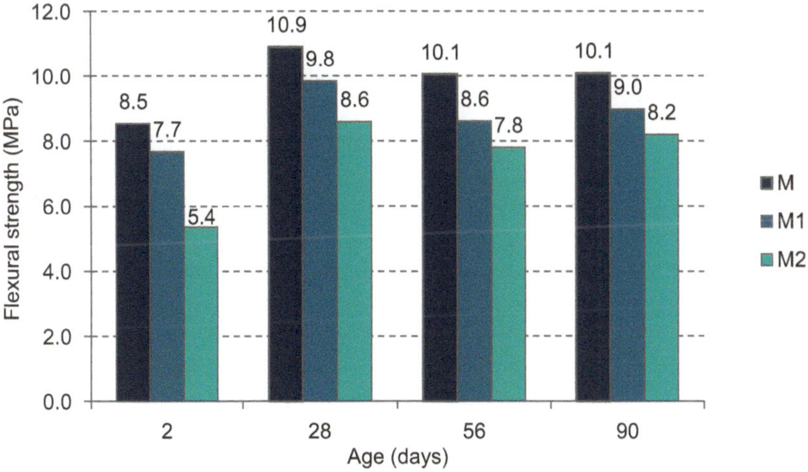

Figure 6.21 Flexural strength development of concrete [4]

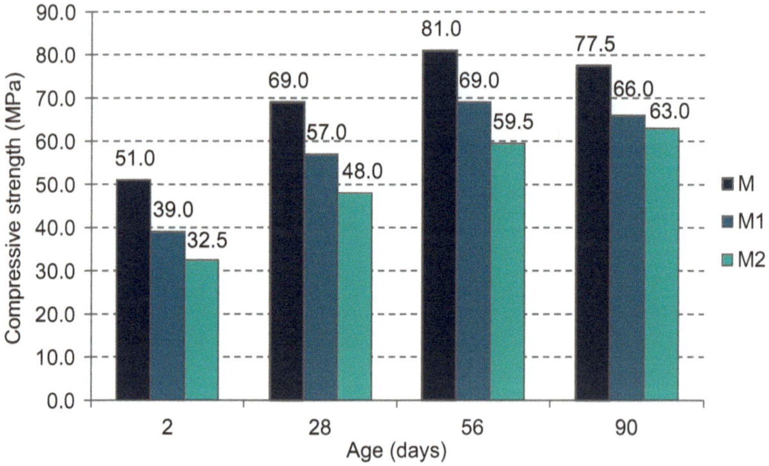

Figure 6.22 Compressive strength development of concrete [4]

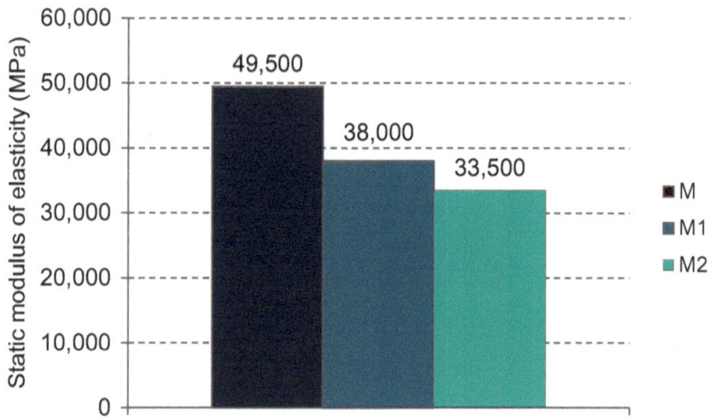

Figure 6.23 Static modulus of elasticity at 56 days of age [4]

of flexural strength of slag concrete compared to the same values for the reference concrete are in accordance with the values of flexural strength shown in [15].

The compressive strength (Figure 6.22) of mixtures with slag increased more than that of the reference mixture. When 2 days old, the M1 mixture achieved 77% of the compressive strength of the reference mixture, while the M2 mixture achieved 64% of the compressive strength of the M mixture. At the end of the observed period (after 90 days), the M1 and M2 mixtures achieved 86% and 81% of the compressive strength of the M mixture, respectively. Lower values for compressive strength of slag concrete compared to reference concrete conflict with those presented in [15] and [16]. As for the modulus of elasticity (Figure 6.23), the M1 mixture developed 77% of the reference mixture value, while the percentage for the M2 mixture amounted to 67%. As a basis for comparison, the slag and limestone mixtures studied in [16] showed similar values for the modulus of elasticity.

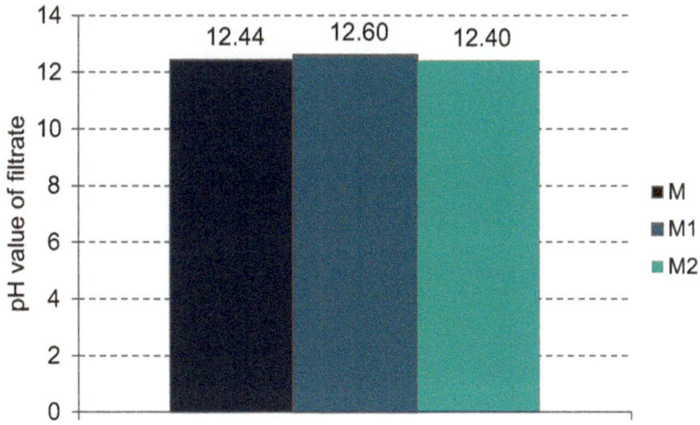

Figure 6.24 The pH values of filtrates [4]

It is obvious that both types of slag yield concrete (M1 and M2) with lower mechanical properties compared to the reference concrete (M) were made with dolomite aggregate. Since all the aggregates were in saturated and surface-dry conditions, lower values for the mechanical properties of slag concrete cannot be explained by the increased absorption of slag as an aggregate. Examination of the fracture lines of the test samples remaining after flexural strength testing revealed that all samples fractured evenly throughout the cement paste and aggregate. Therefore, lower mechanical property values of the slag concrete can be explained by lower quality of slag as an aggregate in comparison with dolomite. However, considering the mechanical properties of the concrete with both types of slag, it may be concluded that these by-products might still be used as a replacement for a commonly used natural aggregate in structural concrete.

Given that all mixtures contained the same volume fraction of aggregates, the filtrate alkalinity (see the pH values shown in Figure 6.24) can be attributed solely to the aggregate. The ratio of alkaline and acidic components in the chemical composition of Slag 1 is $(CaO + MgO)/(SiO_2 + Al_2O_3) = 1.56$, while the same ratio in Slag 2 is 1.78. It would be expected that the Slag 1 filtrate would have a lower pH value than the Slag 2 filtrate; however, it had a higher pH value. Despite the results of the slag pH values being counter to expectations, the slag did not show potential for corrosion here. Specifically, a pH value above 9.5 is considered to be an alkaline medium, in which a passive (protective) layer of iron oxide forms on the surface of the reinforcement. This layer is very dense and impervious to corrosion reactants (O_2, H_2O), and thus it protects the reinforcement from corrosion in a normal, nonaggressive environment [17]. In addition, the pH values of Slag 1 and 2 concrete filtrates differ only slightly from the pH values of the filtrate of the concrete with dolomite. This means that the use of slag as a concrete aggregate is equivalent to the use of dolomite in terms of potential corrosive action.

In addition, susceptibility to corrosion was observed in samples exposed to an aggressive environment for more than 1 year. For long-term exposure to a simulated aggressive environment, concrete "lollipops" with embedded reinforcement were prepared (Figure 6.25). Steel rebar was placed in the centre of concrete specimens,

Figure 6.25 "Lollipops" with embedded reinforcement

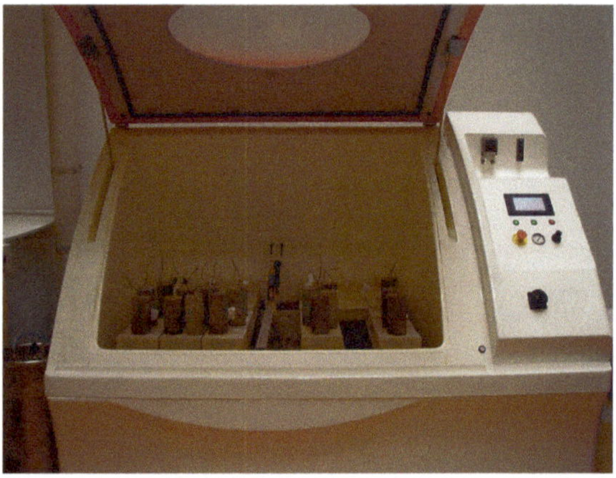

Figure 6.26 Samples exposed to a simulated marine environment in a salt chamber

covered with concrete on all sides. An insulated copper wire was glued on one side of the reinforcing bar to serve as a connection for measuring corrosion parameters. These specimens were left in the humidity chamber for 28 days. After curing, the specimens were placed in a salt spray chamber (Figure 6.26), in which they were exposed to a simulated marine environment with cycles of wetting, drying, and spraying with 3.5% NaCl solution under controlled temperature and humidity conditions. Corrosion parameters (potential and current) and concrete resistivity were periodically measured (Figure 6.27) during the course of exposure by using the galvanostatic pulse method.

From continuous monitoring of the corrosion potential, it was possible to estimate the tendency of steel toward corrosion. Comparing half-cell potential values of samples with dolomite aggregate (M) and those with slag aggregate (M1 and M2), the corrosion potential values of slag aggregate concrete were seen to be significantly lower than those of concrete with dolomite aggregate.

Figure 6.27 Measuring of corrosion parameters

From the corrosion current, the depth of corrosion could be calculated using Faraday's law. The depth of corrosion was defined as the thickness of the reinforcement cross section lost during corrosion.

From these measurements, the loss of reinforcement due to corrosion was found to be similar in concrete made with Slag 1 aggregate and concrete made with dolomite aggregate. In contrast, the corrosion penetration depth with time was higher with the mixture containing Slag 2 aggregate.

Given the pH values of the filtrates of concrete prepared with different types of slag aggregate and the results of corrosion process monitoring during exposure to a simulated aggressive environment in the salt spray chamber, it can be concluded that Slag 1 does not increase the risk of reinforcement corrosion in concrete that will be exposed to an aggressive marine environment. However, Slag 2 produces a negative influence on corrosion. This may indicate an increased risk that corrosion of reinforcement will occur if Slag 2 is used to prepare concrete that will be exposed to an aggressive marine environment. In that case, designing further protection of the concrete is recommended, such as by adding mineral and chemical admixtures or protective coatings.

Figure 6.28 shows the comparison of average values of deformation measured for the reference concrete mixture (M) and mixtures with slags (M1, M2). Each of the curves shown in the figure presents a mean line of 6 measurement lines (3 prisms × 2 measuring lines per prism = 6 measurement lines).

The declining curves indicate the shrinkage of concrete specimens, and the increasing curves indicate their expansion. Although Slag 1 contains a greater amount of free CaO and MgO, which might suggest that the slag mixture would have greater volume changes, no significant difference in volume changes was observed between the Slag 1 and Slag 2 mixtures.

The presented test results do not indicate significantly different volume changes in slag concrete (M1, M2) compared to the reference concrete (M), meaning that the slags did not show an expansive nature. Such a result was expected due to previous

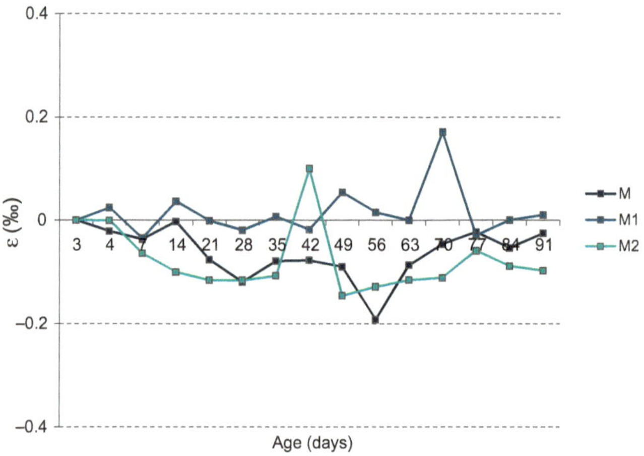

Figure 6.28 Comparison of average values of volume changes [4]

long-term exposure of the studied slags to the influence of atmospheric conditions during the periods that the aggregate was being made from the original slag sample (6 months). During this period, there was enough time for the free CaO and MgO to react. Although the results of the volume stability of the slag mixture are satisfactory in this short reference period, testing of volume stability will continue. If slag shows volume instability, using a method for accelerating slag aging, such as steaming either by itself or under pressure, is recommended [18], which can effectively solve this problem. The use of accelerated aging methods will certainly increase the production costs of slag as an aggregate, but it will also keep it from being a waste material.

The results presented here are affirmative and justify the use of steel slag as an aggregate in structural concrete within Croatia. Additionally, using slag as an aggregate can reduce the exploitation of natural aggregates and thus contribute to environmental preservation.

Because slag is created at high temperatures, it is considered here as an aggregate that could improve the fire resistance of concrete. For this purpose, five concrete mixtures were observed: concrete mixtures given in Table 6.7, as well as concrete mixtures with slag that was previously thermally treated (M1T and M2T). In mixtures M1T and M2T, slag was previously heated to the target temperature of 1000°C at a rate of 5°C/min, and then kept at a constant temperature for 1 h and cooled naturally. The appearance of slag during thermal treatment in a furnace is shown in Figure 6.29.

A number of properties of concrete were tested on 100 × 100 × 400-mm prisms. All specimens were demoulded 24 h after the casting and placed in a water tank for 7 days. Afterward, they were placed in the chamber (at a temperature of about 20±2°C and a humidity level of 95±5%) and kept there until the age of 28 days. At that point, the prisms were removed to the laboratory room until the age of 56 days, at which point the concrete specimens were exposed to high temperatures. The moisture content was 0.6%–0.7% within the samples with thermally treated slag (M1T, M2T) and 3%–4% within the M, M1, and M2 samples. The specimens were put into the test furnace (Figures 6.30 and 6.31), heated to target temperatures of 400°C and 800°C, and then kept at a constant temperature for 1 h. As per RILEM recommendations [19], the

Figure 6.29 Thermal treatment of slag in a furnace

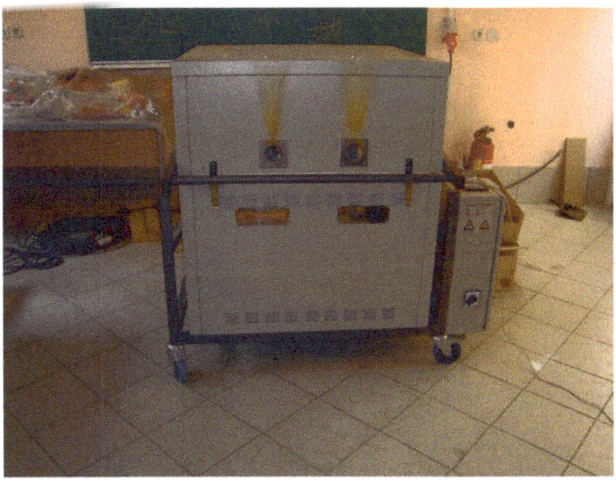

Figure 6.30 The test furnace

Figure 6.31 Samples in the furnace

Figure 6.32 Pulse velocity testing

heating rate was 1°C/min. After the heating process was completed, the furnace was turned off and specimens were allowed to cool inside the furnace to prevent thermal shock. Upon the cooling of the specimens, their mechanical properties (compressive strength and static modulus of elasticity), weight, and pulse velocity were tested (Figure 6.32). The fire resistance of concrete mixtures was estimated by the impact of high temperature on the residual properties of the concrete specimens. The residual value of each property is expressed as the percent reduction of an observed property compared with its initial value in response to an increase in temperature (Figures 6.33–6.40).

In the temperature range from 20°C to 400°C, concrete mixtures with thermally untreated slag (M1, M2) had a smaller reduction in compressive strength compared to mixture M (Figures 6.33 and 6.34). At 800°C, those mixtures lost much more of their compressive strength than did mixture M. Concrete mixtures with thermally treated slag (M1T, M2T) showed superior behaviour at 400°C compared with all other mixtures, even experiencing an increase in compressive strength. However, at 800°C, those mixtures suffered a reduction in strength that was similar to mixtures with thermally untreated slag (M1, M2) and showed no better performance than the reference mixture (M).

As for the modulus of elasticity (Figures 6.35 and 6.36), all mixtures were found to have reduced values with increased temperature. At 400°C, mixtures with slag (except mixture M2) showed better fire resistance than the reference mixture, especially in the case of mixtures prepared with slag that had been thermally treated. Again, the reduction in modulus of elasticity of all mixtures with slag was more pronounced than that of the reference mixture (M) after exposure to 800°C. Compared to thermally untreated slag, thermally treated slag ensured better fire resistance of concrete.

Figures 6.37 and 6.38 show the comparison of the residual weights of concrete prisms with temperature increasing in the furnace, expressed as a percentage of the initial weight (relative weight). In temperatures up to 400°C, the mixtures with thermally untreated slag (M1, M2) lost slightly more weight than the reference mixture (M). Above 400°C, reference concrete experienced a greater reduction in weight than mixtures made of slag (M1, M1T, M2, M2T). This could be attributed to calcium

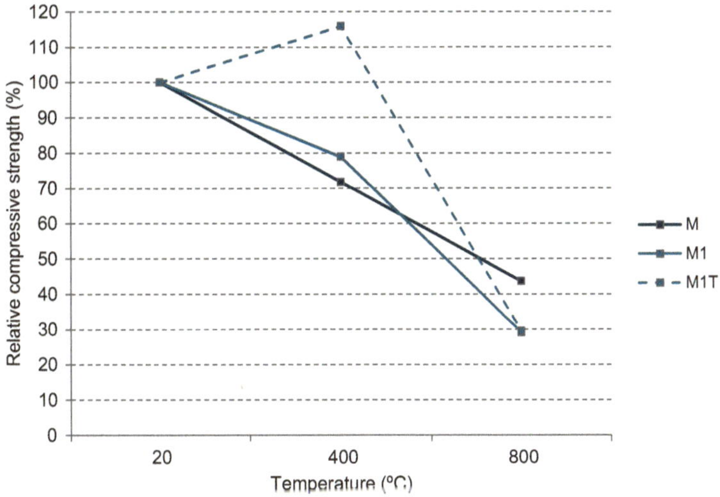

Figure 6.33 Relative compressive strengths of concrete mixtures M, M1, and M1T subjected to high temperatures [20]

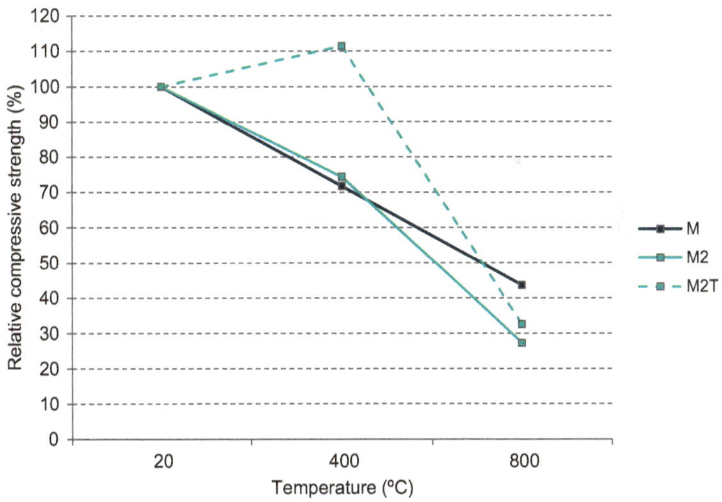

Figure 6.34 Relative compressive strengths of concrete mixtures M, M2, and M2T subjected to high temperatures [20]

carbonate ($CaCO_3$) within the dolomite decomposing into lime (CaO) and carbon dioxide (CO_2), which occurs at a temperature of about 700°C [21-23]. Again, thermally treated slag ensured better fire resistance of concrete than untreated slag throughout the whole temperature range.

In order to estimate the degree of damage within the prisms due to exposure to high temperatures, the pulse velocity was measured. The assumption here was that a significant decrease in velocity with an increase in temperature indicates a greater degree

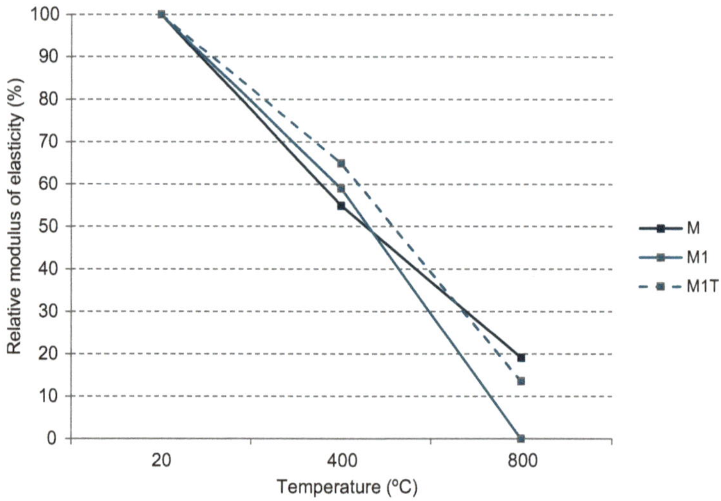

Figure 6.35 Relative modulus of elasticity of concrete mixtures M, M1, and M1T subjected to high temperatures [20]

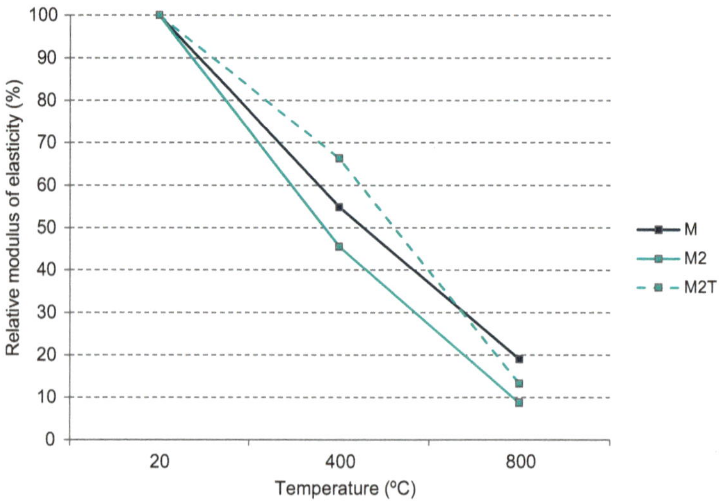

Figure 6.36 Relative modulus of elasticity of concrete mixtures M, M2, and M2T subjected to high temperatures [20]

of damage within the specimen. Figures 6.39 and 6.40 show a comparison of pulse velocities through prisms of all mixtures after exposure to high temperatures. Each result shown represents the mean value of six measurements. Figures 6.39 and 6.40 illustrate that mixtures M1 and M2 experienced a slightly greater reduction in pulse velocity than mixture M for temperatures up to 400°C.

In contrast, mixtures M1T and M2T experienced less reduction in pulse velocity than mixture M. At 800°C, mixtures with thermally treated and untreated slags

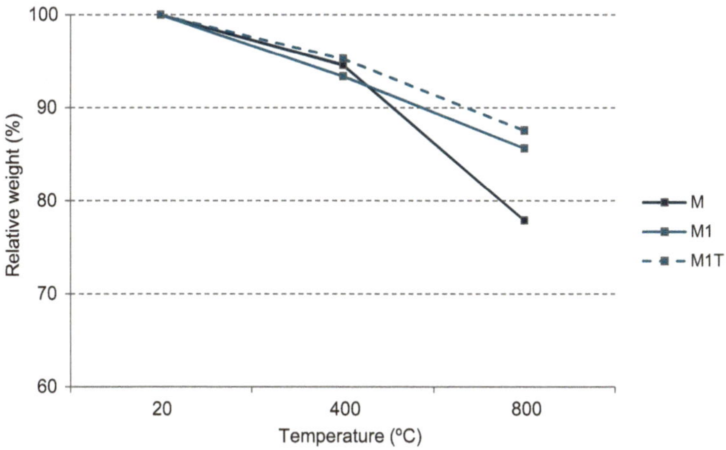

Figure 6.37 Relative weight of concrete mixtures M, M1, and M1T subjected to high temperatures [20]

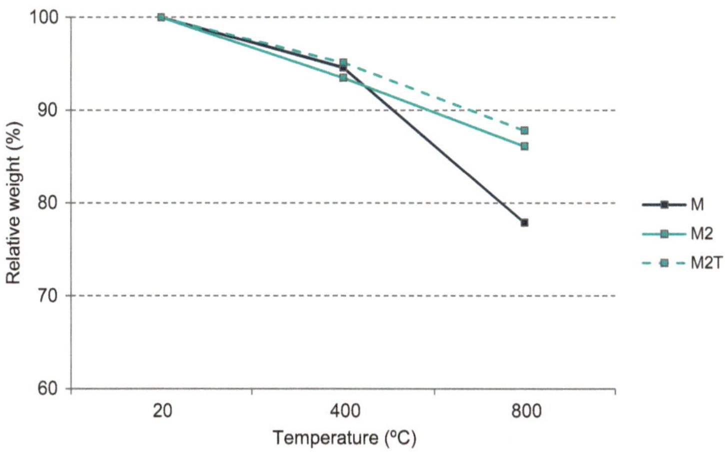

Figure 6.38 Relative weight of concrete mixtures M, M2, and M2T subjected to high temperatures [20]

showed worse fire resistance than mixtures with dolomite (M). This finding leads to the conclusion that a greater degree of damage occurred throughout the entire temperature range within mixtures with thermally untreated slag (M1, M2) compared with the mixture made of dolomite. (The term *damage* includes microcracks that developed during the exposure of specimens to high temperatures.) Specifically, the most commonly used aggregates experience expansion during heating, while cement paste experiences shrinkage at the same time [23]. Differences in thermal behaviour between aggregate and cement paste result in the buildup of parasitic stresses that can lead to the weakening and disruption of the concrete at high temperatures [23, 24]. Thermal treatment of slag diminished that damage and ensured a smaller degree of

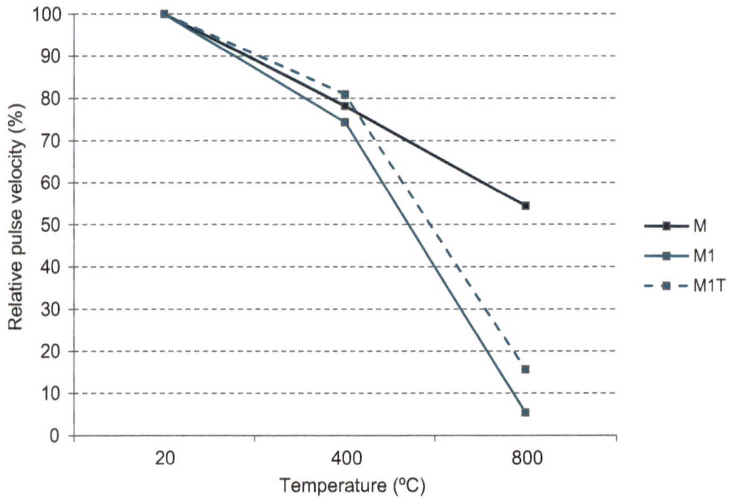

Figure 6.39 Relative pulse velocity of concrete mixtures M, M1, and M1T subjected to high temperatures [20]

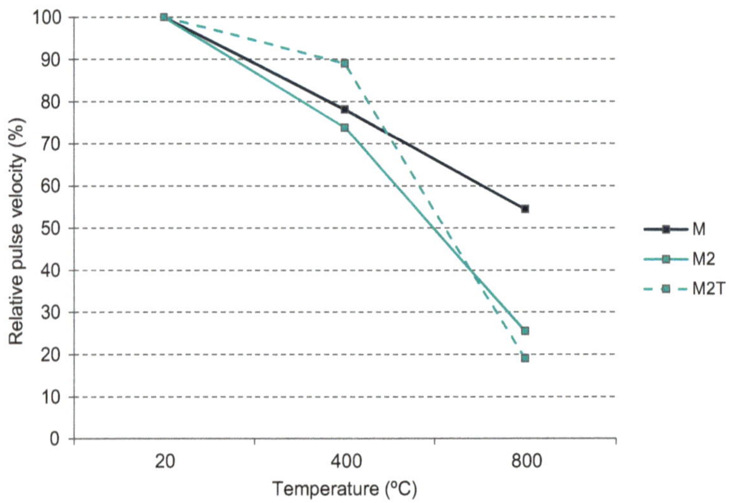

Figure 6.40 Relative pulse velocity of concrete mixtures M, M2, and M2T subjected to high temperatures [20]

damage to concrete with mixtures M1T and M2T in the lower temperature range than to concrete with dolomite (mixture M).

The results of microstructural analysis of concrete specimens (Figure 6.41) are presented in Figure 6.42. Concrete specimens with thermally untreated slag (M1, M2) showed cracks of smaller width than concrete specimens with dolomite (M) at 400°C. It seems that specimens with thermally treated slag (M1T, M2T) had an even better appearance as temperature increases. At 800°C, the cracks had a greater width for all

Figure 6.41 Samples for microstructural analysis

mixtures with slag (M1, M1T, M2, M2T) than for the reference mixture (M). However, using thermally treated slag instead of thermally untreated slag improved the concrete's appearance.

To clarify the obtained results, a dilatometer analysis of cement paste and all aggregates used in the concrete mixtures was conducted (Figure 6.43). Cement paste (C) was made of CEM I 52,5N with a w/c ratio of 0.43. The superplasticiser was not used in the cement paste preparation because it was assumed that it does not affect the fire resistance of mixtures. A dilatometer analysis of all aggregates used in concrete mixtures (dolomite–D, Slag 1–S1, Slag 2–S2, thermally treated Slag 1–S1T, and thermally treated Slag 2–S2T) was also conducted. The samples of cement paste in 40 × 40 × 160-mm prisms were cured in the same way as the concrete samples. At the age of 56 days, samples of 5 × 5 × 20 mm in size were cut out of these prisms for the dilatometer analysis. The results of analysis are shown in Figures 6.44 and 6.45.

The decline of the curve indicates the shrinkage of the sample, while the rising curve indicates expansion. Given that more damage was recorded for mixtures with thermally untreated slag than for the reference mixture, it seemed that the difference in the expansion of aggregate and the shrinkage of cement paste was expressed more in mixtures prepared with thermally untreated slag than with dolomite. Contrary to these expectations, the results of the dilatometric analysis indicated a smaller difference in the behaviour of both slags when untreated (S1, S2) with cement paste (C) than in the behaviour of dolomite (D) with cement paste (C) during exposure to high temperatures. However, it is apparent that thermally untreated slags were unstable throughout the temperature range, changing their behaviour from expansion to shrinkage. Thermal treatment successfully stabilised both slags (S1T, S2T) and ensured their linear expansion during the heating. Such behaviour can be explained by transformation of the mineral composition [25]. Namely, as slag is heated, wüstite transforms to a new mineral, magnetite. This transformation is irreversible and makes slag stable at elevated temperatures. This means that if slag was previously heated to 1000°C, it remains stable after subsequent heating.

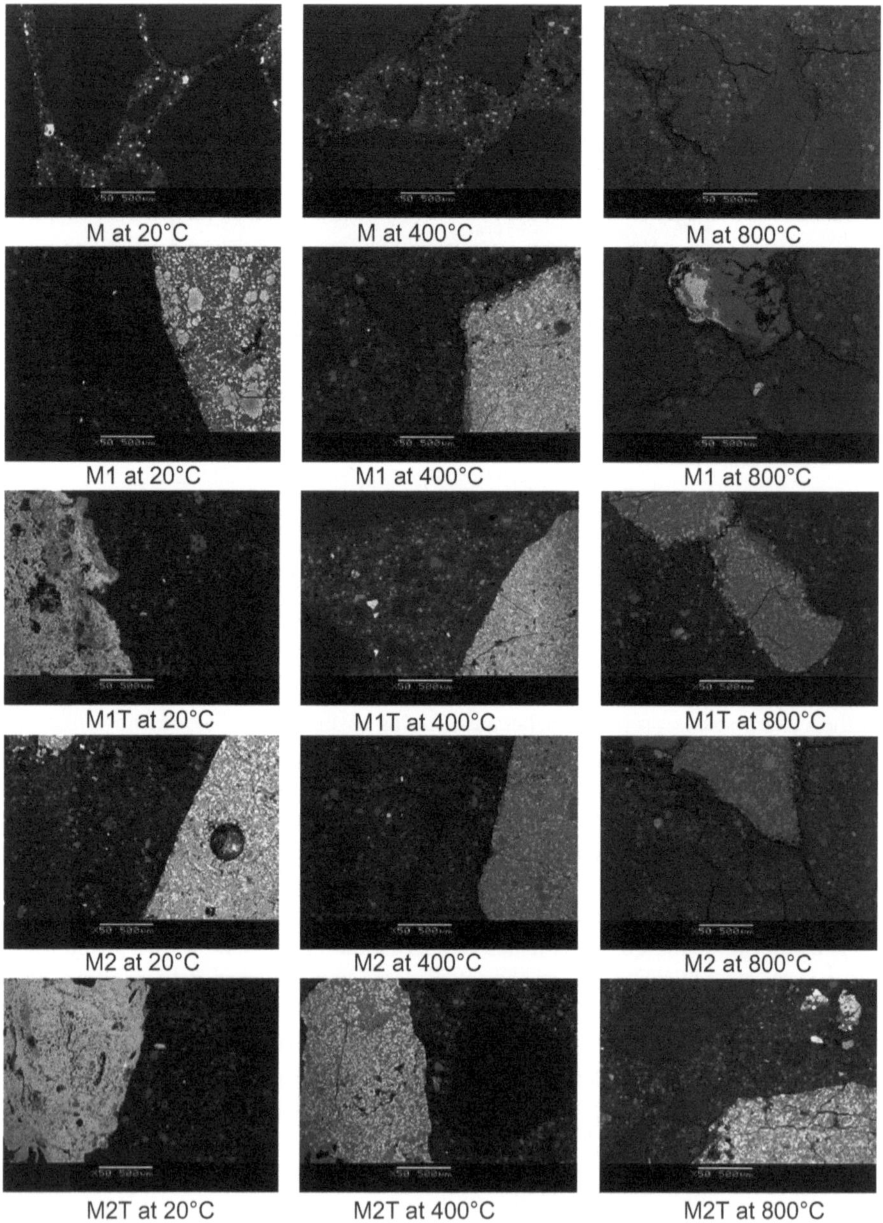

Figure 6.42 Microstructural analysis of concrete mixtures at different temperatures [20]

The results showed that slag could improve the fire resistance of concrete at a lower temperature range (up to 400 °C). This improvement is especially prominent with previously thermally treated slag. However, thermal treatment of slag increases the price of slag concrete, making it economically unacceptable for widespread use.

Figure 6.43 Samples for dilatometer analysis

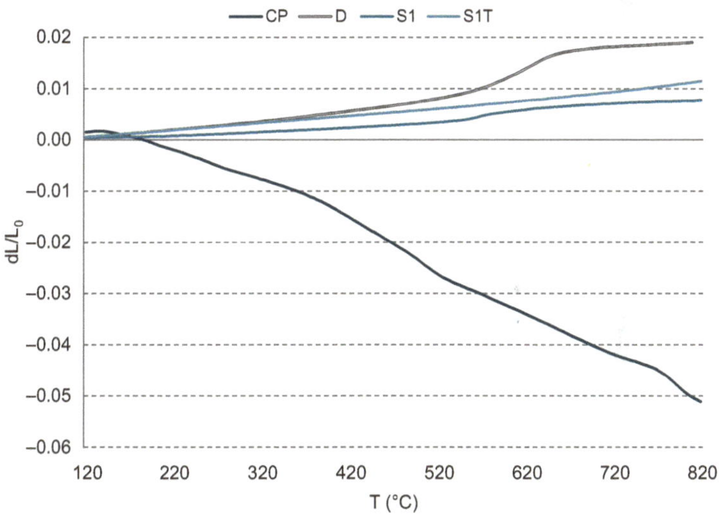

Figure 6.44 The results of dilatometer analysis for cement paste, dolomite, Slag 1 and thermally treated Slag 1 [20]

6.2 Slag usage in road construction

In addition to testing the application of slag in construction, slag from Croatian landfills can be used in road construction for all the layers of pavement structure. The results of studies in the usage of domestic slag in pavement structure (base layers and wearing course) are described next.

6.2.1 Slag usage in unbound layers

Research results of steel slag applicability in unbound pavement base layers are shown in Table 6.9. This material is suitable for building dams, embankments, and unbound

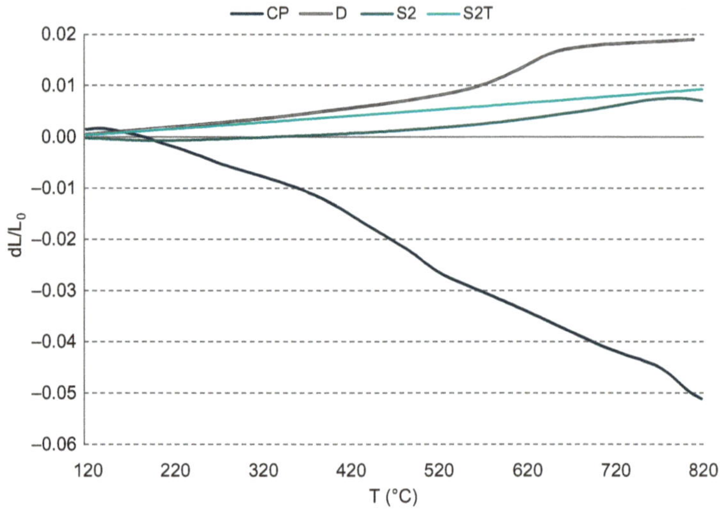

Figure 6.45 The results of dilatometer analysis for cement paste, dolomite, Slag 2, and thermally treated Slag 2 [20]

Table 6.9 Properties of steel slag as material for unbound base layers [26]

Analysed property	Property value	Standard requirement
Density (g/cm^3)	3.831	–
Degree of nonuniformity	6.58	> 4
California bearing ratio (CBR; %)	198	Min. 80
Water absorption, mass %	0.8	Max. 1.6
Frost resistance (Na_2SO_4), mass (%)	0.05	Max. 7
Share of weak and weathered grain, mass (%)	0.4	Max. 12
Resistance to wear by Los Angeles method, mass (%)	17.4	Max. 45

pavement base layers. Tests have also shown that steel slag does not contain organic matter, and there is no danger of pollution even though the pavement structure is in direct contact with the ground.

6.2.2 Slag usage in hydraulically bound base layers

Tests have also shown that steel slag is an equally good substitute for gravel or crushed stone aggregate in both stabilised cement and unbound base layers. Properties of a cement-stabilised mixture of slag, natural crushed stone, and 0.7% and 2% of cement (Class 42.5 N) meet the compressive strength criteria set by Croatian standards [27].

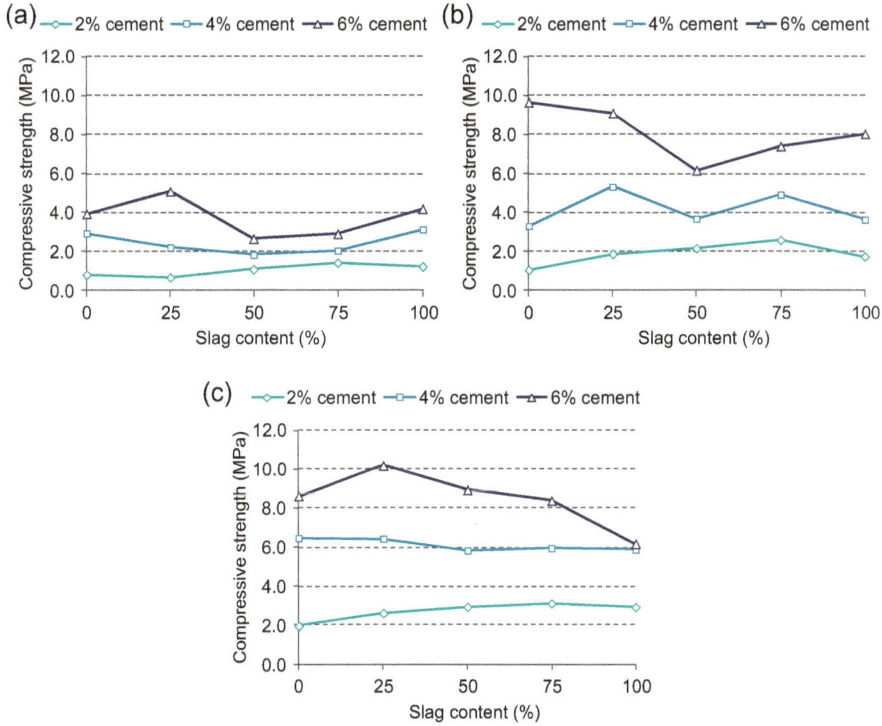

Figure 6.46 Compressive strength for curing ages of (a) 7 days, (b) 28 days, and (c) 90 days [28]

Even with application of gravel as aggregate and lower-strength-class cement [CEM II B/M (PS) 32.5 R], it is still possible to obtain a mixture that meets the strength criteria [28]. In mixes with slag aggregate, there are increases in compressive and indirect tensile strength in mixes with 2% cement content. However, the values generally decrease in mixes with higher cement content (4% and 6%) with increasing slag content. These results are presented in Figures 6.46 and 6.47.

The amount of binder used tends to greatly affect the compressive strength of mixtures containing slag. Filling the surface voids of slag grains with cement paste causes an inconsistent improvement in strength with all mixtures. In Figure 6.48, microscopic images of gravel grain (a) and slag grain (b) are presented. It is clear that gravel grain is covered by cement paste, while the slag grain surface pores are filled by this paste. This phenomenon is shown in Figure 6.49, which presents the relationship between the compressive strength of mixtures with no slag and with 75% slag for various amounts of binder and all curing ages.

As the cement content increases, the difference in compressive strength between mixtures with and without slag decreases. At 6% cement, the compressive strength of the mixture with slag becomes less than that of the mixture without slag. Mixtures with 2% cement and 75% slag had a greater compressive strength than mixtures without slag for all curing ages.

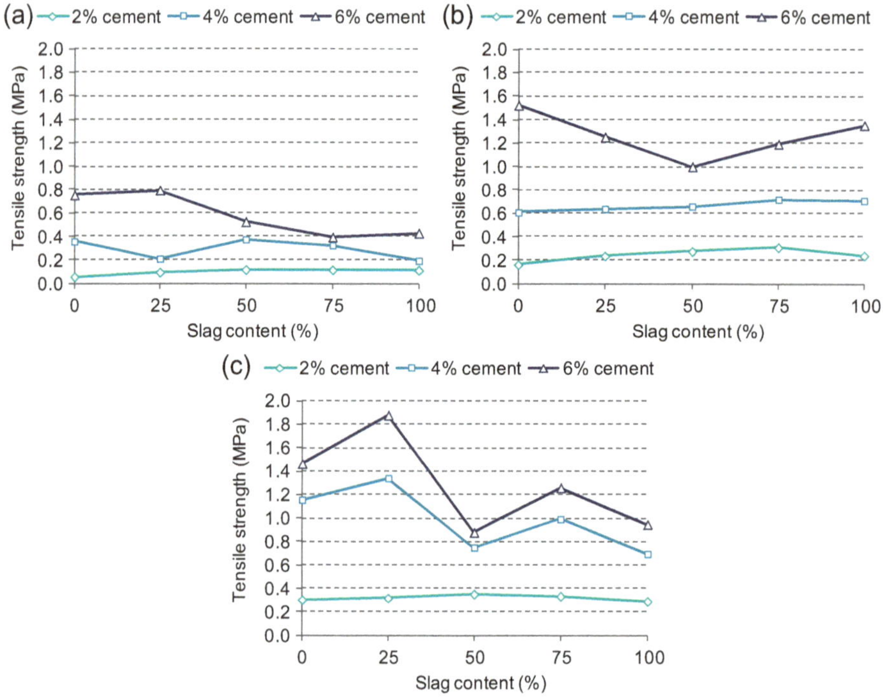

Figure 6.47 Indirect tensile strength for curing ages of (a) 7 days, (b) 28 days, and (c) 90 days [28]

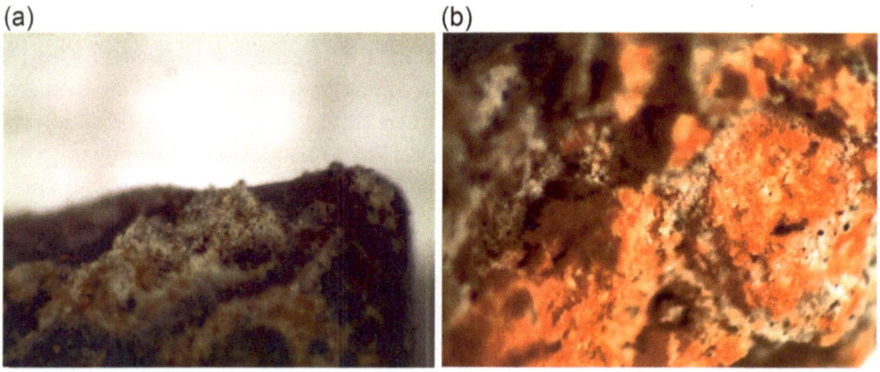

Figure 6.48 (a) Gravel grain and (b) slag grain within a stabilised mixture specimen

Based on EN 14227-1 [29], hydraulically bound mixes are classified according to 28-day compressive strength and sorted into strength classes. All mixes with slag and 2% cement fulfil this criterion for strength class C1.5/2.0. Mixes with 4% cement can generally be classified as strength class C3/4, while mixes with 6% cement cannot be reliably classified into one strength class because the cement content greatly affects the strength development in mixes containing steel slag.

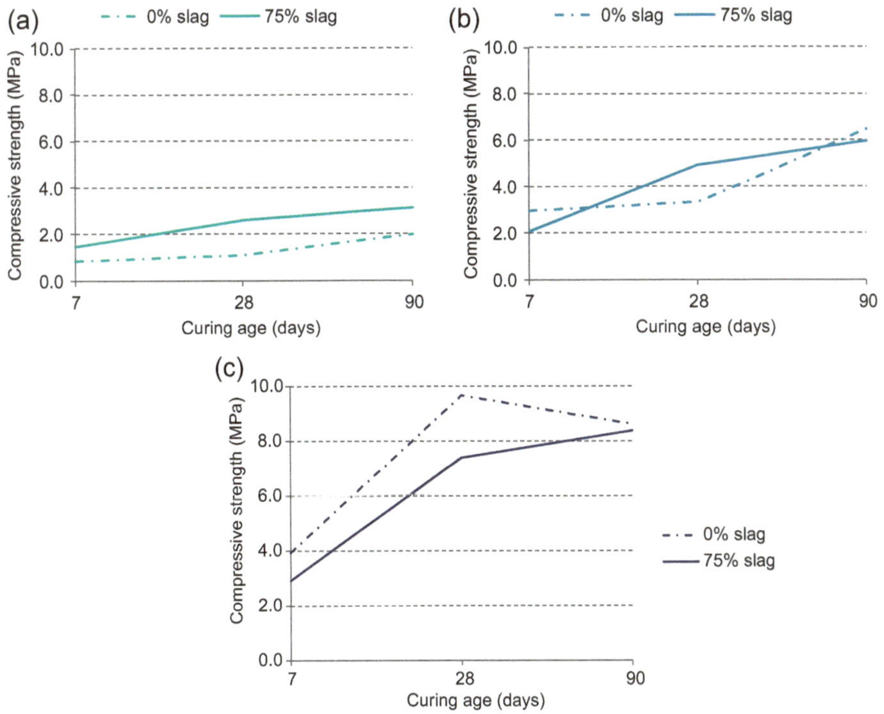

Figure 6.49 Influence of cement content on compressive for cement contents of (a) 2%, (b) 4%, and (c) 6% [28]

During the same study, the modulus of elasticity (static and dynamic) was also investigated. The standard test for determining the modulus of elasticity in hydraulically bound mixtures is defined by EN 13286-43 [30]. The results of this test were very inconsistent, varying greatly among samples within identical mixtures. For example, the static modulus of a mixture with only gravel aggregate varied between 8000 and 18,000 MPa. Therefore, even though this static testing method is given by the European standard for characterising cement-stabilised materials, there are some problems.

First, its accuracy depends on the rate of loading, deformation effects, and the stress level at which the material's elastic properties are determined [31]. Cement-stabilised materials are porous, and the forces applied during the static test crush the material, changing its porosity by an amount that depends on the magnitude of the applied loading. These inconsistencies mean that the accuracy of the measured elastic properties is questionable. Second, directly evaluating the static modulus of elasticity is complicated and time consuming [32]. These complications occur because cement-stabilised materials usually contain an aggregate of gravel or crushed rock. These aggregates create a specimen surface that is very textured and rough, which causes problems with positioning linear variable differential transformers (LVDTs) during measurements (Figure 6.50). The positioning of the LVDTs must be precise when determining the modulus of elasticity and Poisson ratio, so similar variation was found in the results for static modulus as those for the determination of Poisson's ratio.

Figure 6.50 Sample with a mounted LVDT, demonstrating the difficulty of its positioning, and thus the questionable validity of the obtained results [33]

In light of the results and the identified problems, it was decided to conduct further investigation through the measurement of dynamic modulus via the ultrasound pulse velocity and static modulus via the spatial digital image correlation method, namely three-dimensional (3D) digital image correlation (3D DIC). Ultrasound pulse velocity measurement, defined in EN 12504-4 [34] is a standard test method for concrete characterisation.

Cement-stabilised materials, also known as *controlled low-strength materials* or *lean concrete,* have a similar composition as regular concrete. Because of this similarity, the properties of cement-stabilised mixtures were determined by measuring ultrasound pulse velocity, even though it is not a standard test in road construction. For method adjustment, the frequency of 54 kHz was used. For comparison, ultrasonic pulses at frequencies of 40–80 kHz can be used for concrete, while impulses at frequencies of 20–250 kHz can be used for cement-stabilised materials. The test setup for this measurement is presented in Figure 6.51. The dynamic modulus of elasticity can be calculated using Eq. (6.1), with an assumed value of Poisson's ratio equal to 0.25:

$$\mathrm{E} = \rho \cdot \mathrm{v}^2 \cdot \frac{(1+\nu)\cdot(1-2\cdot\nu)}{1-\nu} \tag{6.1}$$

DIC is a nondestructive and noncontact method used to determine a loaded object surface deformation. The fundamental operating procedure is to compare images taken under different loading conditions, which is achieved by dividing the images into small subareas and tracking the movement of these subareas in comparison to the zero position (i.e., before a load is applied). The use of this technique, therefore, enables the measurement of an area's full field deformation, which makes the technique useful for determining the collapse mechanism on various material specimens.

With an insight into the whole deformation field, it is possible to completely reconstruct the crack geometry without extensive instrumentation [36]. Due to the coarse surface of specimens used in this research, a 3D DIC measurement is preferred over

Figure 6.51 Ultrasonic pulse measurement of cement-stabilised steel slag mixtures [35]

Figure 6.52 Test setup for 3D DIC measurement [33]

classic, mechanical devices that are contact based. The testing itself is no different than with mechanical extensometers since the process of recording images can be automated. Image recording is done simultaneously with the acquisition of the analog inputs from the test machine (e.g., force and displacement signal). System software recognises surface patterns on digital images of the measuring object and allocates coordinates to image pixels. Spatial DIC measurement for the purpose of this study was implemented using Aramis 3D optical deformation analysis system, developed by Gesellschaft für optische Messtechnik (GOM). Figure 6.52 shows the Aramis main hardware components, together with the rest of the test setup. The results of this research are presented in Table 6.10.

The results of moduli determined by ultrasound pulse velocity (UPV) measurement, depending on steel slag and cement content, are presented in Figure 6.53. Although there is a weaker correlation between the trend lines and experimental data sets, the equations given in Figure 6.53 can be used to estimate dynamic modulus from slag content, with a known value of dynamic modulus of a mixture without slag. The

Table 6.10 Experimental program results of modulus of elasticity measurement [33]

Share in total aggregate content		Cement in aggregate mass	Density	UPV	Compressive strength	Elastic moduli (UPV)	Elastic moduli (3D DIC)
Gravel	Steel slag						
(%)	(%)	(%)	ρ (kg/m^3)	v (μm/s)	f_c (MPa)	E_{din} (GPa)	E_{st} (GPa)
100	0	2	2,19	2,278	1,38	9,55	0,936
75	25	2	2,30	2,504	1,35	12,01	0,975
25	75	2	2,56	2,485	2,72	13,17	1,334
100	0	4	2,25	3,055	2,99	17,55	3,617
75	25	4	2,34	2,923	2,78	16,69	2,706
25	75	4	2,53	3,081	4,54	20,00	3,650

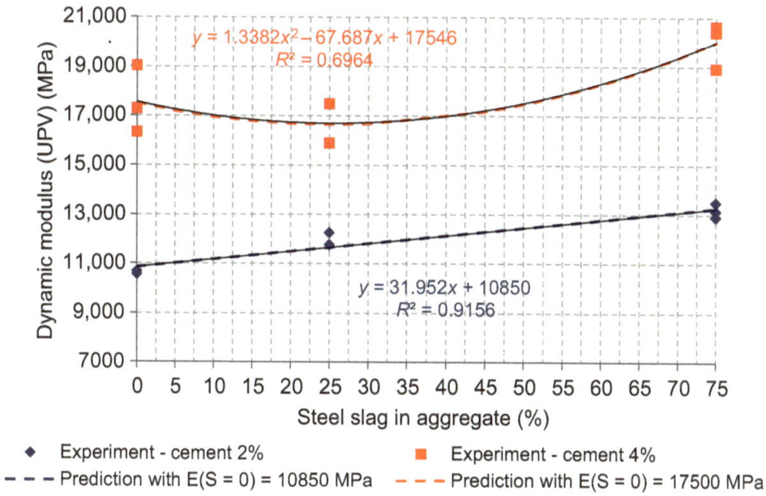

Figure 6.53 Modulus of elasticity determined by UPV, depending on steel slag and cement content [33]

phenomenon of modulus increase with cement content, independent of slag content, is expected due to an increase in compression strength. The same can be said about dynamic modulus increase with slag percentage when cement content is omitted. The equation shown in Figure 6.53 may be considered valid for the investigated mixtures, with *S* expressed in terms of percentage and *E* in MPa. It should be noted that the R^2 value of the shown trendlines would be greater if the mean results of individual mixtures were used instead of the results for each tested sample.

The mixture with 4% of cement in aggregate mass and 75% of slag in aggregate composition has the highest modulus value (E_{din} = 20.00 GPa), and the lowest value is attained by a mixture that has 2% cement content without slag in aggregate composition (E_{din} = 9.36 GPa). These results are in accordance with those found in the literature, which range from 1–20 GPa, depending on the materials' properties [37,38].

The primary advantage of using the 3D DIC method is the possibility of monitoring stress and strain in real time and presenting those results in a clear manner, as shown in Figure 6.54. The EN 13286-43 standard [30] provides guidelines in which the elastic modulus in compression is calculated using a force equal to 30% of the maximum force sustained, and a corresponding strain. Therefore, Figure 6.53 depicts the strain fields at 0%, 30%, maximum, and failure stresses. In addition, the provided visualisation enables the quantification of fracture properties in CSM and monitoring of crack progress. It can be seen that the crack formulation process is heterogeneous and that there are areas in the field with local deformations even below the 30% stress limit.

The results of static modulus of elasticity measurements by 3D DIC are presented in Figure 6.55. They range from 0.9–3.5 GPa, which is in accordance with data found in literature, ranging from 1.5–3 GPa, depending on the cement content [39]. When analysing the static modulus results shown in Figure 6.55, the trends established between groups of mixtures with 2% and 4% of cement content are very similar to those established for the dynamic modulus, shown in Figure 6.53. The increase of

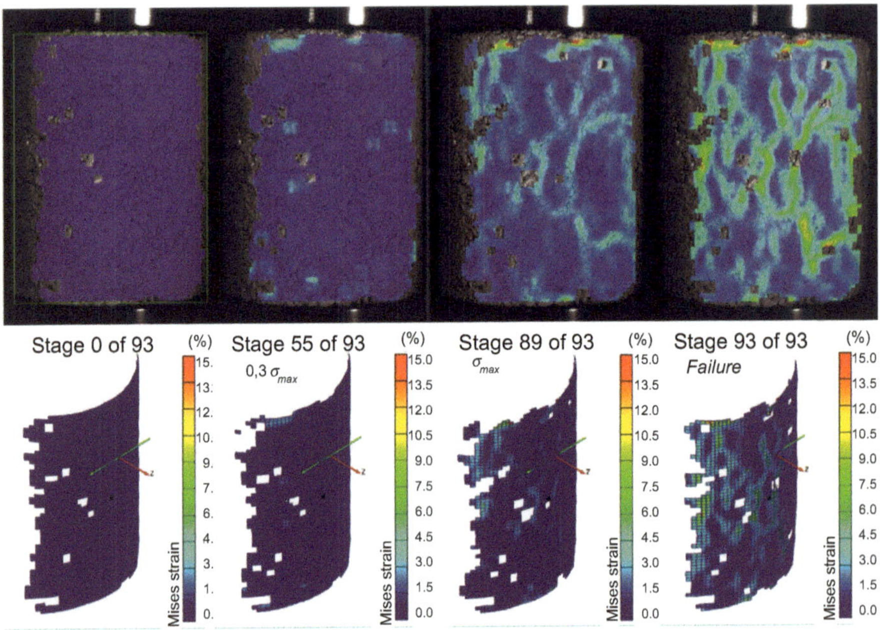

Figure 6.54 Visualisation of the strain state during compressive testing [33]

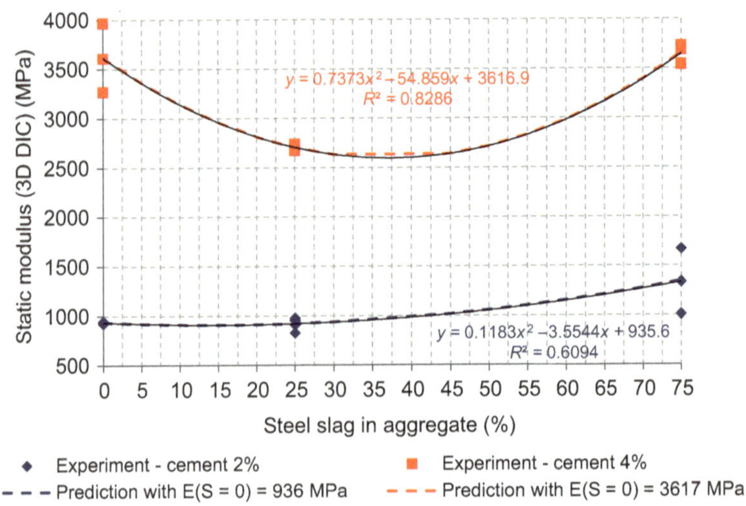

Figure 6.55 Modulus of elasticity determined by 3D DIC technique, in comparison with steel slag content [33]

modulus with slag and cement content is the most obvious conclusion to reach. The highest modulus was recorded for the mixture that contained 75% slag aggregate and 4% cement content (3.65 GPa), and the lowest value was recorded for the mixture composed entirely of gravel and 2% cement content (0.93 GPa). The beneficial effect of slag is more pronounced in mixtures with 2% cement. Mixtures with 4% cement and 75% slag attain the same values of modulus as a 100% gravel mixture, while the other percentages of slag (Figure 6.55) result in lower values of modulus. With cement content of 2%, the modulus value is constantly increasing with the percentage of slag in aggregate.

6.2.3 Slag use in wearing courses

6.2.3.1 Slag use in asphalt mixtures

Following trends with research worldwide, domestic slag also has been used in asphalt mixtures as the most common form of its application. By replacing the fractions 4–8 and 8–16 mm of natural aggregate with slag from the Split landfill and fillers (stone dust) by electrofilter ash from a power plant in Tuzla and amorphous silicon dust, an asphalt mixture of increased density and stability was produced [40]. The results of this study are shown in Table 6.11.

Tests related to the application of steel slag from another Croatian landfill (i.e., the Sisak landfill) were also performed. This research resulted in the construction of a trail section as an example of good practice.

To test the suitability of steel slag from the Sisak landfill for its application in the manufacture of asphalt mixtures, analyses commonly used to assess natural aggregate for the same purpose were conducted [41]. The results showed better physical and mechanical properties of steel slag in comparison with natural carbonate aggregate. The trail section was constructed in 2009 during the reconstruction of a county road. The width of the test field was 6 m, the length was 200 m, and the thickness of asphalt layer was 4 cm. In one

Table 6.11 Physical and mechanical properties of asphalt mixture with waste material

Analysed property	AC-11s	AC (slag-Split)	AC (fly ash)	AC (SiO_2 dust)
Stability by 60°C (kN)	11.10	12.40	15.30	12.20
Density by 25°C (kg/m^3)	2619	2713	2528	2510
Voids content (%)	4.36	4.73	4.37	4.73
Voids filled with binder (%)	75.60	77.60	75.86	74.31
Density of stone mixture (kg/m^3)	2883	3066	2782	2761
Share of bitumen (vol.%)	5.48	6.44	8.36	8.77

Table 6.12 Physical and mechanical properties of asphalt mixture with steel slag

Analysed property	AC 11 (Sisak steel slag aggregate)	AC 11 (carbonate aggregate)
Soluble binder content, % (m/m)	5.1	5.4
Bulk density (kg/m^3)	2724	2368
Maximum density (kg/m^3)	2927	2469
Void content, % (v/v)	6.9	4.1
Void in the mineral aggregate, % (v/v)	20.6	16.6
Void filled with binder, % (v/v)	66.3	75.4
Stability (kN)	15.2	11.7
Tangential flow (mm)	1.1	0.9
Flow (mm)	2.5	1.9
Marshall quotient (kN/mm)	6.1	6.2
Mass of dry sample (g)	1249.7	1243.5
Water sensitivity of bituminous specimens (kPa), at 25°	1608	1513
Water sensitivity of bituminous specimens (kPa), at 40°	1287	1215
Indirect tensile strength ratio (%)	80.0	80.3
Indirect tensile strength of bituminous specimens (kPa), at 5°	3988	3638
Indirect tensile strength of bituminous specimens (kPa), at 25°	1847	1700
Stiffness modulus (MPa)	7128	5642

lane, the worn existing course was replaced with an AC 11 type of asphalt with natural aggregate, and the other lane with AC 11 with 70% steel slag aggregate. The results (shown in Table 6.12) showed good physical and mechanical properties of the asphalt mixture with steel slag in comparison with the natural aggregate mixture.

During this investigation, no problems were encountered in the design or the placement of asphalt with steel slag. Steel slag asphalt for the surface course had excellent wheel tracking and skid resistance and good wear resistance [41].

6.2.3.2 *Slag use in pervious concrete*

Pervious concrete, also referred to as *porous* or *permeable concrete,* is a material with the same basic components as standard concrete, but it is designed to have high porosity (void content is between 11% and 35% [42]) and permeability (typically about 2–6 mm/s). It contains a mixture of Portland cement, uniform coarse aggregate, either a small amount of fine aggregate or none, and water [43]. Due to its high porosity, pervious concrete has good drainage properties and high noise absorption characteristics, which are important elements for high-quality pavement. Its drainage properties have enabled its use in the construction of shoulders, bases, and subbases of roads. More recently, it has been used in the construction of the top layer or overlay of concrete pavement due to its good acoustic properties, which are similar to those of porous asphalt [42]. The high porosity of pervious concrete is also its main disadvantage since that is associated with a decrease in strength. The low strength of pervious concrete influences the stability and durability of a structure because of the susceptibility of the concrete to frost damage and its low resistance to damage caused by chemicals. These factors explain its limited application in the construction of high-traffic highways. In order to address these issues, studies on new components for pervious concrete are being conducted worldwide [43-45].

In order to verify the possibility of using Croatian slag in pervious concrete, five different concrete mixtures were prepared. This study was intended to determine the optimal mixture for pervious concrete to meet the requirements of Croatian legislation concerning pavement [46]. According to the General Technical Conditions for Roadwork [46], concrete for pavement is classified based on its 28-day compressive and flexural strengths. The required compressive strength for a very heavy traffic load is C 35/45, while flexural strength must be 5.0 MPa. For a heavy traffic load, the required compressive strength is C 30/37 and flexural strength is 4.5 MPa, while the required load compressive strength and flexural strength for other traffic are C 25/30 and 4.0 MPa, respectively. The first mixture in this study was a standard dense concrete mixture, while the others were pervious concrete mixtures with two types of aggregates and proportions of aggregate fractions (Table 6.13). Crushed dolomite stone was used as a natural material, and steel slag from the landfill near the town of Sisak was used as a substitute aggregate material.

Pervious concrete mixtures M2 and M4 were prepared from coarse dolomite aggregate (with fractions of 4–8 mm and 8–16 mm) and sand from the Drava River (with fractions of 0–2 mm); the mixtures M3 and M5 were prepared from a coarse steel slag aggregate (with fractions of 4–8 mm and 8–16 mm) and sand from the Drava River (with fractions of 0–2 mm); and the reference mixture (M1) was prepared entirely from dolomite (with fractions of 0–4 mm, 4–8 mm, and 8–16 mm). Each pervious concrete mixture contained 10% sand from the Drava River. The grain size distribution of all aggregate fractions is shown in Figure 6.56, and the cumulative sieving curves of aggregates in specific concrete mixtures are shown in Figure 6.57. The grain size distribution of the aggregates was determined according to EN 933-1:2012 [47]. The designations M1–M5 in Figure 6.57 correspond to concrete mixtures M1–M5.

The dolomite or steel slag aggregates in mixtures were used with different proportions of finer and coarser fractions (with ratios of 60:30 and 30:60). For all mixtures, the w/c ratio was 0.33. The cement was ordinary Portland cement, CEM II/A-M(S-V) 42.5 N, according to EN 197-1:2011 [49], and with a density of 3.0 kg/dm^3, according to EN 196-6:2010 [50]. The cement content was 300 kg/m^3 for all mixtures except the reference mixture (M1). Fractions of aggregates were mixed first for 3 min with the cement in a pan mixer (DZ 100VS, Diemwerke). Water was subsequently added, and mixing continued for another 5 min.

The consistency was measured by slump testing according to EN 12350-2:2009 [9]. For all mixtures, it was 0 or 1 cm. Three specimens were prepared to determine the

Table 6.13 Mixture compositions [48]

Characteristics	Mixture									
	M1		M2		M3		M4		M5	
Water/cement proportion (w/c)	0.33		0.33		0.33		0.33		0.33	
Cement (kg)	350.0		300.0		300.0		300.0		300.0	
Water (kg)	115.5		99.0		99.0		99.0		99.0	
Superplasticiser (kg)	3.5		0.0		0.0		0.0		0.0	
Aggregate (kg)	2034.0		1783.7		2053.3		1783.7		2053.3	
Components										
Sand, 0–2 mm (%/kg)	—	—	10	178.4	10	205.3	10	178.4	10	205.3
Dolomite, 0–4 mm (%/kg)	40	813.6	—	—	—	—	—	—	—	—
Dolomite, 4–8 mm (%/kg)	30	610.2	60	1070.2	—	—	30	535.1	—	—
Dolomite, 8–16 mm (%/kg)	30	610.2	30	535.1	—	—	60	1070.2	—	—
Steel slag, 4–8 mm (%/kg)	—	—	—	—	60	1232.0	—	—	30	616.0
Steel slag, 8–16 mm (%/kg)	—	—	—	—	30	616.0	—	—	60	1232.0
Total (kg)	2503.0		2182.7		2452.3		2182.7		2452.3	

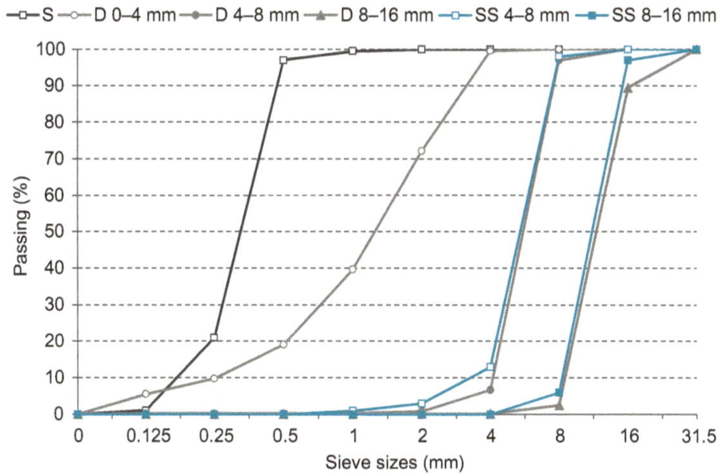

Figure 6.56 Sieving curves of aggregate fractions (S = sand, D = dolomite, SS = steel slag) [48]

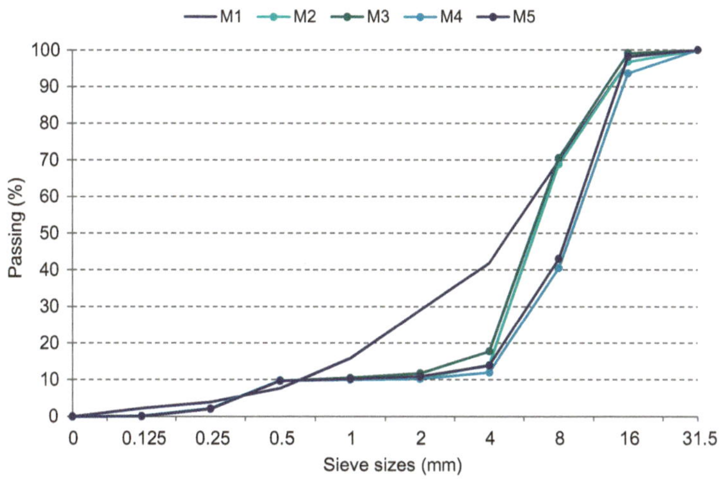

Figure 6.57 Cumulative sieving curves of aggregates in concrete mixtures [48]

individual mechanical properties. Specimens of pervious concrete were compacted with a compacting rod and specimens of reference mixture by a vibrating table. All specimens were extracted from the moulds 24 h after the casting and placed in a water tank for 27 days at a temperature of 20°C ± 5°C in accordance with EN 12390-2:2009 [51].

At 28 days, the mechanical properties of the hardened concrete specimens were tested according to the relevant European standards. Compressive strength was tested on cube specimens of 15-cm edge length with a constant rate of loading of 0.5 MPa/s, according to EN 12390-3:2009 [12]. Density was tested on the same specimens using the principles described in EN 12390-7:2009, and the volume of the specimens was determined by water displacement [52]. The flexural strength of prism specimens measuring 10 × 10 × 40 cm^3 was tested by loading them at a constant rate of 0.05 MPa/s according to EN 12390-5:2009 [10]. Dynamic modulus of elasticity was tested on prism specimens with the same measurements, using ultrasonic pulse velocity according to EN 12504-4:2004 [34]. Poisson's ratio was selected to be 0.22 [53].

The total porosity was determined through the following steps: (1) determination of the dry mass and saturated surface-dried mass of the specimen (weighed in water and air); (2) calculation of density and apparent density from parameters from step 1; and (3) calculation of solid material volume with a basic formula for density and isolated pore volume with a basic formula for apparent density, on the assumption that the total volume of the specimen was a sum of solid material volume, isolated pore volume, and connected pore volume. In addition to this calculation, porosity was determined by Avizo Fire 3D image analysis software, which is based on two-dimensional (2D) X-ray images (threshold segmentation method and characterisation of different pore spaces have been written about [54,55]. The structure of individual specimens (a 4.5-cm-diameter cylinder for all specimens, ranging from 5 to 6 cm in height depending on the specimen) was investigated by X-ray computed microtomography (MicroXCT-400, Xradia). The X-ray energy was set to values of 140 kV and 125 μA for all samples using the macro 0.39× objective (i.e., the field of view was between 17.0 and 50.0 mm), and 4000 projection images were taken from different viewpoints; the pixel resolution was 34–46 μm, depending on the specimen. The connectivity of pores was defined as the ratio of volume of connected pore space and volume of total pore space [56].

Table 6.14 shows the results of the hardened concrete tests. The pervious concrete (M2–M5) had less compressive and flexural strength but more porosity and lower density than the reference mixture (M1), as expected.

As shown in Table 6.14, the total porosity in pervious concrete ranged from 14%–22%, but pervious concrete with dolomite aggregate (concrete mixtures M2 and M4) had greater porosity (by about 23%) than pervious concrete with steel slag aggregate (M3 and M5). A higher proportion of the coarse fraction in mixtures (M4 and M5) with both types of aggregate (dolomite and slag) resulted in greater total porosity, which agreed with observations of the American Concrete Institute [57].

Table 6.14 Results of hardened concrete tests [48]

	Mixture				
Characteristics	**M1**	**M2**	**M3**	**M4**	**M5**
Compressive strength (MPa)	69.50	21.90	26.50	26.60	20.20
Flexural strength (MPa)	9.70	4.90	4.00	4.00	2.80
Density (kg/m^3)	2442.4	2095.0	2435.4	2076.6	2337.8
Total porosity (%)	6.30	21.29	14.24	22.20	19.27
Dynamic modulus of elasticity (GPa)	70.85	51.27	37.61	52.04	33.70

Figure 6.58 Fractured pervious concrete, M4 (a) and M5 (b) [48]

Mixtures with slag (M3 and M5) showed greater density compared to mixtures with dolomite (M2 and M4), which resulted from the greater density of the slag as an aggregate. The higher density of concrete mixtures was observed in mixtures with a higher amount of small aggregate fractions (M2 and M3).

Concrete mixtures with dolomite (M2 and M4) had more flexural strength and dynamic modulus of elasticity than did mixtures with slag (M3 and M5). Flexural strength increased with a greater amount of small aggregate fraction (M2 and M3). The influence of aggregate type and size on compressive strength, as well as its influence on dynamic modulus of elasticity, were not observed.

Pieces of fractured specimens are shown in Figure 6.58. Contrary to [58], the influences of aggregate type and size were not observed here, and all specimens fractured in the same way, cracking evenly through their structure.

The results of pore space characteristics of pervious concretes obtained by microtomography are summarised in Table 6.15, which presents the total volume of solid material and pore space, volume of connected pore space (pores that are connected to others and to the external surface), and volume of isolated pore space (pores that are not connected to the external surface). These values show that more total porosity was obtained in M2 and M4 (21.4% and 21.2%, respectively), but specimen M4 reached the greatest open (connected) porosity (Figure 6.58), which is crucial for the permeability of pervious concrete.

Compared to specimens M3 and M5, in which the same proportions of material were used, only the aggregate was different (slag instead of dolomite) in M2 and M4. It is worth noting that with the use of slag, both the overall porosity and the open porosity decreased significantly. It can be seen from Figures 6.59 and 6.60 in the case of slag (M3, M5) that total porosity and connected porosity were lower than the dolomite aggregate samples (M2, M4). Since the porosity obtained by hydrostatic weighing did not differ to a great extent among samples, samples with slag may possess capillary pores that were not detected by microtomography because of the resolution. But the

Table 6.15 Summary of pore space characteristics of pervious concrete obtained by microtomography [48]

	Mixture				
Characteristics	**M1**	**M2**	**M3**	**M4**	**M5**
Total volume of solid material (mm^3)	21,704.4	27,761.7	44,926.3	33,916.6	39,975.1
Volume of total pore space (mm^3)	813.2	7565.8	7095.3	9143.3	4857.8
Volume of connected pore space (mm^3)	163.1	4054.3	3478.4	6638.5	2883.2
Volume of isolated pore space (mm^3)	650.1	3511.5	3616.9	2504.7	1974.6
Total porosity (%)	3.6	21.4	13.6	21.2	19.81
Connected porosity (%)	0.7	11.5	6.7	15.4	13.92
Isolated porosity (%)	2.9	9.9	6.9	5.8	5.89
Connectivity of pores	0.2	0.5	0.5	0.7	0.7

main parameter for characterising pervious concrete is pore connectivity, which was the highest in sample M4.

The microstructure and chemical composition of the polished cross sections of the pervious concrete specimens were examined by the backscattered electron image mode of low-vacuum scanning electron microscopy (SEM), using JEOL 5500 LV equipment. SEM analyses were performed to examine the interfacial zone between the cement paste and both types of aggregates, dolomite and slag. It is known from previous studies [59,60] that the contact zone between the cement paste and the aggregate is the system's weakest point. This is partially due to the different morphology of the cement paste at the boundary with the aggregate, but also due to water being able to accumulate beneath the large grains, where the local w/c ratio is higher. The next influential parameter on the contact zone is physical interaction—that is, if the aggregate is rough and porous, then the cement paste penetrates the aggregate and forms a stronger bond [61].

As can be seen in Figures 6.61(a–b), the cement paste was homogenous with both aggregate types. Further, there was no difference (e.g., densification) in the cement paste at the aggregate boundary. The interlocking effect of the cement matrix with the slag aggregate can be clearly seen where the cement paste entered the open porosity on the boundary of grain of the aggregate [Figure 6.61(b)]. This was not the case when dolomite was used as the aggregate [Figure 6.61(a)]. Based on the results, the following conclusions can be made:

- More total porosity is obtained in pervious concrete with dolomite aggregate compared to pervious concrete with steel slag aggregate.
- A higher proportion of coarse fraction in mixtures results in greater overall porosity.
- Surprisingly, connected porosity as a main parameter for the estimation of pervious concrete efficiency is influenced more by the aggregate's type than its size. The microtomographic investigation showed that in the case of slag aggregate, both overall and open porosity are lower.

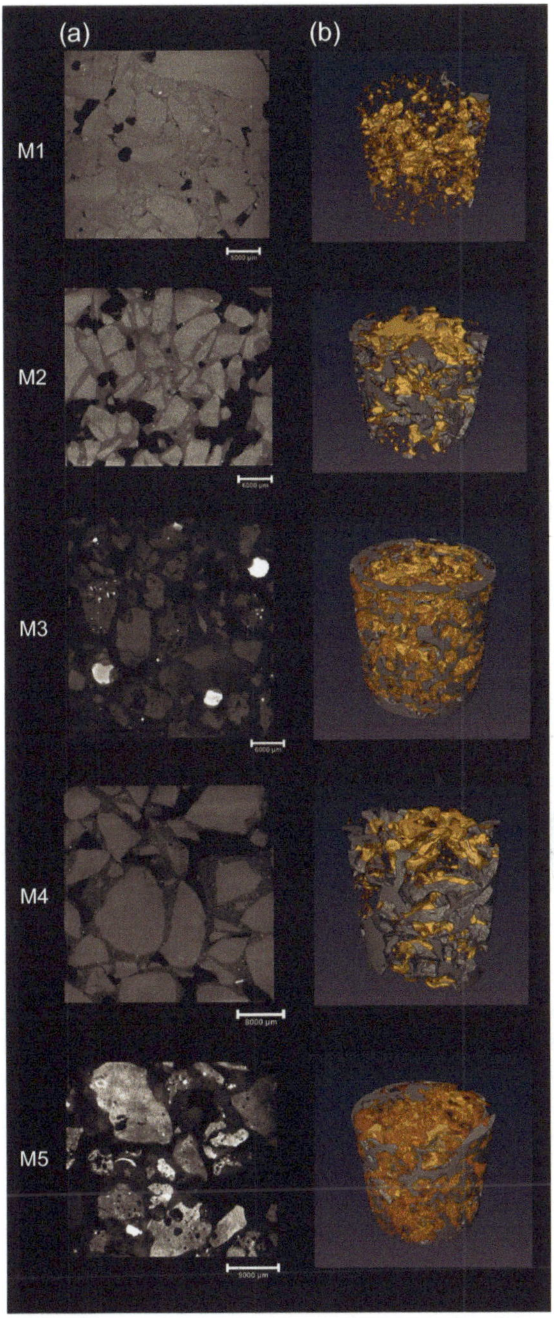

Figure 6.59 X-ray microtomography images of samples M1 to M5; (a) detail of the XZ slice image and (b) 3D pore shape (gray indicates connected pore space; yellow indicates isolated pore space) [48]

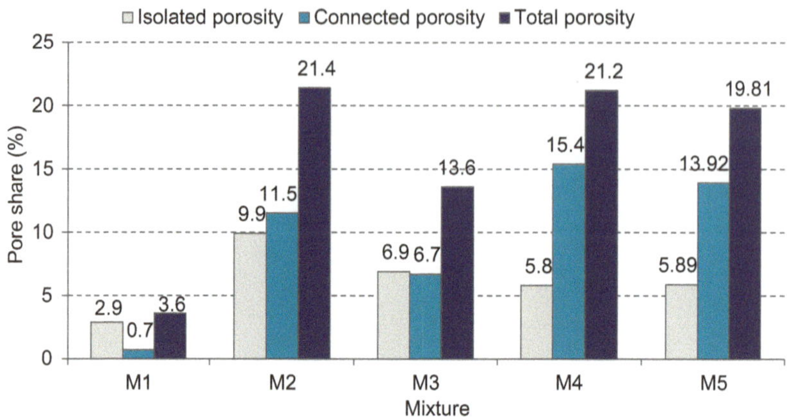

Figure 6.60 Distribution of porosity among the specimens obtained by the Avizo program from Figure 6.59 [48]

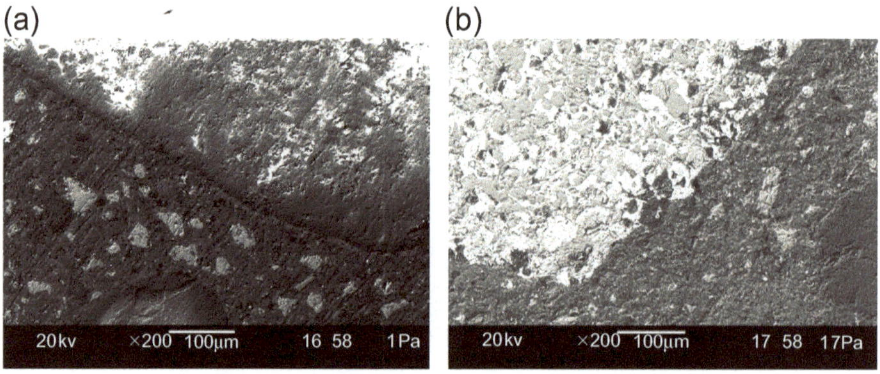

Figure 6.61 Backscattered electron SEM image of (a) M2 and (b) M3 [48]

- In the case of slag aggregate, the interlocking effect of the cement matrix with the slag aggregate is noticed because the cement paste penetrated into the open porosity of the slag aggregate. This interlocking effect might contribute to the higher mechanical strength.
- A higher quantity of small aggregate fractions (4–8 mm) influenced the greater density and flexural strength of the concrete mixtures.
- Pervious concrete does not satisfy the requirements for pavement concrete for 28-day strength according to General Technical Conditions for Roadwork [46], but it can be used as concrete block paving.

6.3 Environmental aspects of slag usage in road construction

In order to give complete insight into the possibility of using slag in unbound road layers or embankments through a multidisciplinary approach to its environmental impact, leaching testing has been conducted as one of the standard test methods in assessing the environmental influence of slag. However, it has also been tested on earthworms to evaluate its impact on creatures directly exposed to slag influence when it is embedded in road embankment. Also, radioactivity measurement has been performed since it is known that slag may contain certain amounts of radionuclides such as ^{226}Ra, ^{232}Th and ^{40}K [62,63]. All those measurements have been conducted within the COST Action TU1301 – Norm4Building project, supported by the EU Framework Programme.

The leachability test was conducted according to EN 12457-2. One stage batch test at a liquid-to-solid ratio of 10/1 l/kg for materials with particle sizes below 4 mm (with or without size reduction) [64]. This procedure uses distilled/deionised water as the leachant, with a contact time of 24 h. The liquid-to-solid ratio is 10 L/kg using material with a particle size below 4 mm. Although in this research [65], slag is treated like coarse aggregate in road construction, 4-mm particles are tested since in general, leachability is assumed to increase when particle size decreases [66]. Reduction of the particle size will decrease the resistance against diffusion of substances out of the particle and increase the rate of release of substances that are not otherwise affected, resulting in less favourable conditions. The test results are presented in Table 6.16. Besides leaching tests, the total concentration of heavy metals in steel slag is determined in order to compare the available content with its leaching potential. The total heavy metal concentration is determined according to standards ISO 11466 and ISO 11047 [67,68]. In order to determine the possible adverse steel slag impact on the environment, total element concentration and released concentration are compared. To do this, Eq. 6.2 [66] is used to convert mg/l of released element (content in leachate) to mg/kg, taking into account the liquid-to-solid ratio used in leachate preparation.

$$\begin{aligned} Subtance\ release\ (mg/kg) = {} & Subtanceconcentration\ in\ eluate\ (mg/l) \\ & \times L/S\ ratio\ (l/kg) \end{aligned} \tag{6.2}$$

Analysing the results of the total heavy metal content in two steel slag samples presented in Table 6.16, high values of the Ca, Mg, Cr, and Fe oxides are observed, which was expected according to the slag chemical composition presented earlier in this chapter. However, the content of these heavy metals in leachate is very low, ranging from 0% (Ca, Mg, and Fe) to 0,02% (Cr). The reason why the elements are kept from leaching at levels proportional to their content lies in the material's microstructure. The concentration of heavy metal in leachate depends on the way that element is incorporated into the aggregate matrix, which can make it inaccessible for contact with water (leachant) [66]. So, even under the conditions of size reduction to a fine granular material (as in this research [65]) or acidic conditions, the available heavy metal fraction for leaching may be a very minor portion of its total content.

On the other hand, even though the total concentration of V-pentaoxide is considerably less than in Cr (242.12 mg/kg and 10563.00 mg/kg, respectively), vanadium content in leachate is the highest of all tested heavy metals (0,078% for Slag 1 and 0,037% for Slag 2). However, comparing the results of heavy metal concentration in leachate and limit values for surface wastewater discharges according to Croatian regulations [which are in accordance with the 2000/60/EC, 2006/118/EC, and 2006/11/EC

Table 6.16 **Heavy metal concentrations in slag samples [65]**

Component	Slag 1			Slag 2		
	Total content (mg/kg)	Leaching (mg/L)	% Leached element	Total content (mg/kg)	Leaching (mg/L)	% Leached element
Al	7033.33	0.0685	0.010	2970.00	0.0792	0.027
Ca	106,000.00	0.0173	0.000	64,066	0.0526	0.001
Mg	131200.00	0.061	0.000	38,666	0.0643	0.002
Cr	10,563.00	0.221	0.021	7780.00	0.015	0.002
Ti	222.36	Not measured	–	388,33	Not measured	–
Mn	1778.00	Not measured	–	581,00	Not measured	–
Fe	9230.00	0.071	0.008	7153.33	0.0526	0.007
V-pentaoxide	242.12	0.0189	0.078	92.50	0.0034	0.037

Table 6.17 Heavy metal emission limit values for surface wastewater discharges [73]

Element	Maximal limited value (mg/L)
Al	3
Cr–total	0.5
Cr (VI)	0.1
Fe	2
V	0.05
Cu	0.5
Zn	2
Mn	2
Hg	0,01
Fe	2

directives [69–71] (Table 6.17)], it can be concluded that tested steel slag does not present environmental risk in terms of leaching heavy metals. Even though trace amounts of vanadium are essential for normal cell growth, higher concentration of this element can be toxic [72]. Therefore, when steel slag is used in road construction, vanadium concentration must be monitored.

With the same sample of leachate prepared to determine heavy metal concentration, the pH value and dissolved oxygen (DO) are measured as well [65]. Namely, the toxicity and bioavailability of the metal contaminants to terrestrial organisms depend on soil properties, such as soil pH (in soil with low pH, many metals are more mobile and available for uptake than in neutral or alkaline soil) [74]. According to the Croatian regulation, the pH value of surface wastewater is limited to 6.5–9 [73]. The DO level in water (or leachate, in this case) is an important environmental factor indicating water quality since oxygen is necessary to all forms of life. It has a great influence on the lives of some sensitive animals and plants, which may relocate, weaken, or die if its level is too low or too high. Total dissolved oxygen concentrations in water exceeding 25 mg/L can be harmful to aquatic life (especially for eggs and small larvae, and some fish and aquatic invertebrates may suffer from gas bubble disease) and if the DO level drops below 5.0 mg/L, aquatic life is put under stress [75,76]. DO levels fluctuate seasonally over a 24-h period and vary with water temperature and altitude: cold water holds more oxygen than warm water, and water holds less oxygen at higher altitudes. Also, for high levels of DO, corrosion of steel pipes can accelerate [75].

For this test, Multimeter 18.52.01 Eijkelkamp was used. The measurements were conducted at room temperature (24.3°C). The equipment and testing for DO and pH value determination are shown in Figure 6.62, and the results are presented in Table 6.18.

Analysing the results presented in Table 6.18, it can be concluded that both samples of slag leachate have a sufficient amount of DO within the range $5\ mg/L < DO < 25\ mg/L$. This result contributes to the finding that the steel slag tested in this research does not present potential risk to the environment [65]. Based on the obtained results (DO

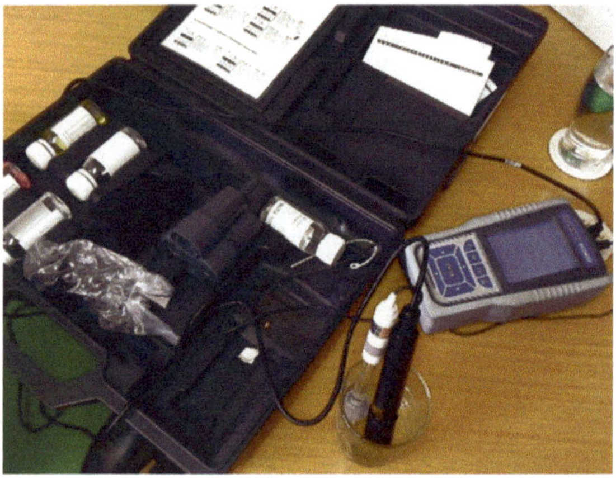

Figure 6.62 Leachate pH and DO measurement [65]

Table 6.18 DO and pH values of leachate for two slag samples [65]

Property	Slag 1	Slag 2
DO (%)	67.8	64.4
DO (mg/L)	5.66	5.36
pH	9.68	9.41

levels of 5.66 mg/L and 5.36 mg/L for Slag 1 and Slag 2, respectively), there is no risk of faster corrosion of steel pipes if steel slag is used in embankments.

Analysing the leachate pH values, slightly alkaline leachate is observed with pH values of 9.68 and 9.41 for Slag 1 and Slag 2, respectively, which is slightly more than the limit value of 9. This alkaline leachate is the result of the dissolution of Ca silicates, oxides, and carbonates [77]. This can have an impact on the environment in terms of calcareous crystalline crust formation, similar to natural travertine deposits. This phenomena can smother benthic and littoral aquatic habitats, reduce light penetration to benthic primary producers, high pH values can be harmful to fish populations and increase chemical oxygen demand, high sulphate loadings, salinity, and amphoteric or anionic heavy metals (e.g., arsenic and selenium) [77,78]. However, leachate from tested slag samples is just slightly over the limit value of 9; therefore, it is expected that the alkaline leachate will be neutralised through interaction with the acidic soil [79].

Furthermore, in order to investigate the impact of steel slag as a road construction material, mortality and molecular biomarkers were measured on earthworms exposed to the extracts of two steel slag samples [65]. To prepare the extracts, 200 g of material and 200 ml of distilled water were placed in the flask and then in a rotary shaker for 24 h. Thus, prepared extracts were exposed to earthworms using contact filter paper testing. The contact filter paper test is the acute toxicity test, which is performed according to OECD 207 protocol [80]. The principle of the test is to expose individual

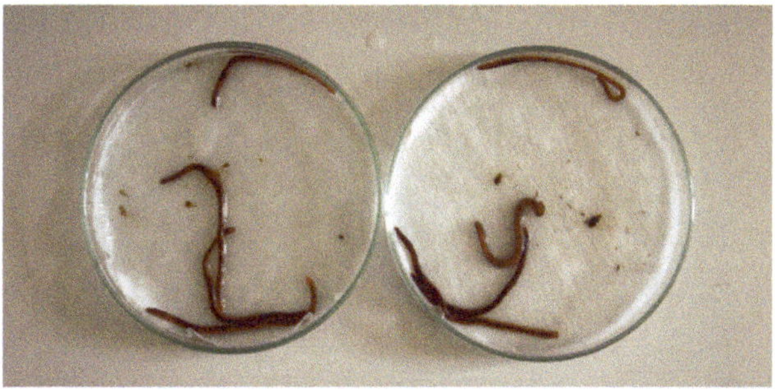

Figure 6.63 Earthworms exposed to slag samples using the filter paper contact test [65]

Table 6.19 Measured and calculated values of radionuclides in two slag samples [65]

Radionuclide	Activity concentration (Bq/kg)	
	Slag 1	Slag 2
^{238}U	27,9 ± 2.6	24,9 ± 2.7
^{232}Th	13,3 ± 0.4	6,3 ± 0.4
^{235}U	1,5 ± 0.1	1,0 ± 0.1
^{226}Ra	22,6 ± 0.8	17,1 ± 0.8
^{210}Pb	20,3 ± 1.2	30,8 ± 1.3
^{40}K	82,5 ± 0.9	105,3 ± 1.0
^{7}Be	1.8 ± 0.3	1.8 ± 0.4
^{137}Cs	4.0 ± 0.1	7.1 ± 0.1

earthworms to the test substance in glass plates whose surfaces are coated with a filter paper, and the test substance is distributed over the surface filter paper (Figure 6.63).

Earthworms are exposed to different dilutions (10%, 25%, 50%, 75%, and 100%) of two slag samples for 48 h (Figure 6.63). The control group (0%) contained distilled water rather than the extract. The results showed that exposure to extracts of both slag samples after 48 h did not lead to mortality in the earthworms.

After the period of exposure, earthworms were homogenised and the homogenates were centrifuged (9000 g, 30 min, 4°C) to obtain a supernatant (postmitochondrial fraction, S9) that was used to measure molecular biomarkers. The measurement of molecular markers showed that short-term exposure to different dilutions of tested slag samples did not have adverse effects on earthworms. As the measured enzymes are present in various organisms (including humans), it can be concluded that tested slags would have similar effects on them as well [65].

Table 6.19 shows the measured and calculated values of certain radionuclides in slag samples.

Table 6.20 **Estimation of the possibility to use steel slag as a building material in road construction in terms of radioactivity [65]**

Country	Activity index (Bq/kg)		Activity concentration value	
	Slag 1	Slag 2	Slag 1	Slag 2
Germany	–	–	22.6 < 200 Bq/kg	17.1 < 200 Bq/kg
Finland	0.070 < 1	0.052 < 1	–	–
Latvia	0.120 < 1	0.078 < 1	82.5< 2500 Bq/kg	105.3< 2500 Bq/kg
Lithuania	0.069 < 1	0.050 < 1	–	–
Norway	0.169 < 1	0.124 < 1	22.6 < 200 Bq/kg	17.1 < 200 Bq/kg
Poland	0.141 < 1	0.102 < 1	22.6 < 185 Bq/kg	17.1 < 185 Bq/kg
Luxembourg	–	–	82.5 < 5000 Bq/kg, 13.3 < 250 Bq/kg and 22.6 < 350 Bq/kg	105.3 < 5000 Bq/kg, 6.3 < 250 Bq/kg and 17.1 < 350 Bq/kg
Slovak Republic	46.320<1	34.031<1	–	–
Croatia	0.169<1	0.124<1	82.5 < 300 Bq/kg, 13.3 < 200 Bq/kg and 22.6 < 200 Bq/kg	105.3 < 300 Bq/kg, 6.3 < 200 Bq/kg and 17.1 < 200 Bq/kg

Gamma spectrometric analysis detected the presence of natural radionuclides such as ^{40}K; the members of natural uranium and thorium radioactive series, ^{232}Th and ^{235}U; and ^{7}Be. In addition to these, ^{137}Cs was detected as well. The activity index of each croatian slag is calculated and compared with the maximum allowed values different countries. Also, the activity concentration values of certain radionuclides are compared with the recommended concentrations in building materials intended for use in road structures [65]. A comparison of results regarding this aspect is shown in Table 6.20.

It can be seen from Table 6.20 that the activity indices and activity concentration values for both slags are below the allowed values in each of the countries in the study. In that sense, these slags should not be considered as radioactive waste, and thus there is no reason not to use it as a building material in road construction [65].

The results shown here indicate that the analysed slags present no hazard to the environment if used in road construction (unbound pavement layers or embankments).

6.4 Concluding remarks

The results of the research presented in this chapter concern the possibility of the use of slag in a wide variety of civil engineering areas. However, these are the results of scientific studies that have not yet become part of practical application. Although Croatia encourages the use and reuse of waste declaratively, it does not punish its disposal. Also, during this crisis, both iron factories stopped operating, and, currently, there is no continuous production of slag in Croatia that would make it necessary to achieve uniformity in its quality. Uneven quality is a problem that plagues slag, as well as all other recycled aggregates whose possibility of application in the particular area of civil engineering has been scientifically proven, which interferes with their use in practice.

References

[1] European Committee for Standardization, EN 12620/AC:2006, Aggregates for concretes, CEN/TC 154, Brussels, 2006.

[2] European Committee for Standardization, EN 1744-1:2004: Tests for chemical properties of aggregates, Part 1: Chemical analysis, CEN/TC 154, Brussels, 2004.

[3] Croatian Ministry of Environmental and Nature Protection, Technical requirements for reinforced concrete structures—TPBK (in Croatian), Zagreb: Official Gazette 139/09, 14/10, 125/10, 136/12, 2006.

[4] I. Netinger, D. Bjegović, G. Vrhovac, Utilisation of steel slag as an aggregate in concrete, Mat. Struct. 44 (9) (2011) 1565–1575.

[5] N.G. Surendra, et al., The chemistry of dicalcium silicate mineral, J. Mat. Sci. 14 (1979) 1554–1566.

[6] M. Muravljov, Građevinski materijali, (in Serbian), 4th ed., Građevinska knjiga, Beograd, Yugoslavia, 2000.

[7] European Committee for Standardization, EN 12350-6:2009: Testing fresh concrete, Part 6: Density CEN/TC 104, Brussels, 2009.

[8] European Committee for Standardization, EN 12350-7:2009: Testing fresh concrete, Part 7: Air content—Pressure methods, CEN/TC 104, Brussels, 2009.

[9] European Committee for Standardization, EN 12350-2:2009: Testing fresh concrete, Part 8: Slump test, CEN/TC 104, Brussels, 2009.

[10] European Committee for Standardization, EN 12390-5:2009: Testing hardened concrete, Part 5: Flexural strength of test specimens, CEN/TC 104, Brussels, 2009.

[11] HRN U.M1025, Method for testing modulus of elasticity of concrete (in Croatian), Croatian Standards Institute, 1982.

[12] European Committee for Standardization, EN 12390-3:2009: Testing hardened concrete, Part 3: Determination of compressive strength, CEN/TC 104, Brussels, 2009.

[13] W. Sun, et al., The influence of mineral admixtures on resistance to corrosion of steel bars in green high-performance concrete, Cement Concr. Res. 34 (2004) 1781–1785.

[14] HRN U.M1029, Concrete—Determination of volume changes (in Croatian), Croatian Standards Institute, 1983.

[15] M. Maslehuddin, et al., Comparison of properties of steel slag and crushed limestone aggregate concretes, Const. Bldg. Mat. 17 (2003) 105–112.

[16] M. Etxeberria, et al., Properties of concrete using metallurgical industrial by-products as aggregates, Const. Bldg. Mat. 24 (2010) 1594–1600.

[17] J. Francisković, Corrosion protection in reinforced-concrete structures (in Croatian), Građevinar 50 (1998) 193–202.

[18] N.O.D. Silveira, et al., ACERITA – Steel slag with reduced expansion potential, in: H. Motz, S. Haimi, M. Mäkikyrö (Eds.). 4th European Slag Conference: Slags—Providing Solutions for Global Construction and Other Markets, EUROSLAG Publication, Oulu, Finland, 2005, pp. 145–157.

[19] International Union of Laboratories and Experts in Construction Materials, Systems and Structures RILEM Technical Comity 129-MHT, Test methods for mechanical properties of concrete at high temperatures, Part 3: Compressive strength for service and accident conditions.

[20] I. Netinger, M. Jelčić Rukavina, A. Mladenovič, Improvement of post-fire properties of concrete with steel slag aggregate, Proc. Eng. 62 (2013) 745–753.

[21] O. Arioz, Effects of elevated temperatures on properties of concrete, Fire Safety J. 42 (2007) 516–522.

[22] A. Savva, P. Manita, K.K. Sideris, Influence of elevated temperatures on the mechanical properties of blended cement concretes prepared with limestone and siliceous aggregates, Cement Concr. Comp. 27 (2005) 239–248.

[23] G.A. Khoury, et al., Fib Bulletin 38: Fire Design of Concrete Structures—Materials, Structures, and Modelling, International Federation for Structural Concrete (fib), Lausanne, Switzerland, 2007.

[24] T. Uygunoğlu, İ.B. Topçu, Thermal expansion of self-consolidating normal and lightweight aggregate concrete at elevated temperature, Const. Bldg. Mat. 23 (2009) 3063–3069.

[25] V. Ducman, A. Mladenovič, The potential use of steel slag in refractory concrete, Mat. Char. 62 (2011) 716–723.

[26] A. Rastovčan-Mioč, T. Sofilić, B. Mioč, Application of electric arc furnace slag, in: K. Grilec, G. Marić (Eds.), Matrib 2009, Hrvatsko društvo za materijale i tribologiju, Korčula, Croatia, 2009, pp. 436–444.

[27] I. Androjić, S. Dimter, Compressive strength of steel slag stabilized mixes, Građevinar 64 (1) (2012) 15–21.

[28] I. Barišić, S. Dimter, T. Rukavina, Strength properties of steel slag stabilized mixes, Composites: Part B 58 (2014) 386–391.

[29] European Committee for Standardization, EN 14227-1: Hydraulically bound mixtures—Specifications—Part 1: Cement bound granular mixtures. Brussels, 2004.

[30] European Committee for Standardization, EN 13286-43:2003: Unbound and hydraulically bound mixtures, Test method for the determination of the modulus of elasticity of hydraulically bound mixtures. CEN/TC 227, Brussels, 2003.

[31] O. Bahr, et al., Young's modulus and Poisson's ratio of concrete at high temperatures: Experimental investigations, Mat. Des. 45 (2013) 421–429.

[32] D.K. Panesar, B. Shindman, Elastic properties of self consolidating concrete, Const. Bldg. Mat. 25 (2011) 3334–3344.

[33] I. Barišić, T. Dokšanović, H. Draganić, Characterization of hydraulically bound base materials through digital image correlation, Const. Bldg. Mat. 83 (2015) 299–307.

[34] European Committee for Standardization, EN 12504-4:2004: Testing concrete, Part 4: Determination of ultrasonic pulse velocity, CEN/TC 104, Brussels, 2004.

[35] I. Barišić, Influence of steel slag stabilized mixtures on pavement stress-strain characteristics, in: 3rd International Conference on Advances in Civil and Structural Engineering, Institute of Research Engineers and Doctors, Kuala Lumpur, Malaysia, 2015.

[36] T.M. Fayyada, J.M. Lees, Application of digital image correlation to reinforced concrete fracture, Proc. Mat. Sci. 3 (2014) 1585–1590.

[37] B. Babić, Relationships between mechanical properties of ement stabilized materials, Mat. Struct. 20 (1987) 455–460.

[38] D.X. Xuan, et al., Mechanical properties of cement-treated aggregate material—A review, Mat. Des. 33 (2012) 496–502.

[39] AustStab Technical Note No. 5, Cement stabilisation practice, AustStab Pavement Recycling and Stabilisation Association, Cherrybrook, Australia, March 2012.

[40] M. Mikoč, D. Marković, Influence of slag, fly ash, and silica fume on the mechanical and physical properties of asphalt, Tehnički vijesnik 17 (4) (2010) 505–514.

[41] T. Sofilić, et al., Steel slag application in Croatian asphalt mixture production, 2nd International Scientific Conference Management of technology step to sustainable production (MOTSP 2010): Conference Proceedings, Rovinj, Croatia (2010).

[42] B.J. Putman, A.I. Neptune, Comparison of test specimen preparation techniques for pervious concrete pavements, Const. Bldg. Mat. 25 (2011) 3480–3485.

[43] B. Huang, et al., Laboratory evaluation of permeability and strength of polymer-modified pervious concrete, Const. Bldg. Mat. 24 (2010) 818–823.

[44] M.Á. Pindado, A. Aguado, A. Josa, Fatigue behavior of polymer-modified porous concretes, Cement Concr. Res. 29 (1999) 1077–1083.

[45] B. Gerharz, Pavements on the base of polymer-modified drainage concrete, Coll. Surf. A: Physicochem. Eng. Asp. 152 (1999) 205–209.

[46] Croatian Roads - Croatian Motorway, General Technical Conditions for Roadwork, Concrete Works (in Croatian), Zagreb, Croatia, 2001.

[47] European Committee for Standardization, EN 933-1: Tests for geometrical properties of aggregates—Part 1: Determination of particle size distribution—Sieving method. CEN/TC 154, Brussels, 1997.

[48] K. Ćosić, et al., Influence of aggregate type and size on properties of pervious concrete, Const. Build. Mat. 78 (2015) 69–76.

[49] European Committee for Standardization, EN 197-1: 2011 Cement. Composition, specifications, and conformity criteria for common cements, CEN/TC 51, Brussels, 2011.

[50] European Committee for Standardization, EN 196-6:2010: Methods of testing cement—Part 6: Determination of fineness. CEN/TC 51, Brussels, 2010.

[51] European Committee for Standardization, EN 12390 2:2009: Testing hardened concrete—Part 2: Making and curing specimens for strength tests, CEN/TC 104, Brussels, 2009.

[52] European Committee for Standardization, EN 12390 7:2009: Testing hardened concrete—Part 7: Density of hardened concrete specimens, CEN/TC 104, Brussels, 2009.

[53] H. Wang, Q. Li, Prediction of elastic modulus and Poisson's ratio for unsaturated concrete, Intl. J. Solids Struct. 44 (5) (2007) 1370–1379.

[54] L. Korat, et al., Characterisation of the pore-forming process in lightweight aggregate based on silica sludge by means of X-ray micro-tomography (micro-CT) and mercury intrusion porosimetry (MIP), Ceram. Intl. 39 (6) (2013) 6997–7005.

[55] M. Henry, I.S. Darma, T. Sugiyama, Analysis of the effect of heating and re-curing on the microstructure of high-strength concrete using X-ray CT, Const. Bldg. Mat. 67 (Part A) (2014) 37–46.

[56] G. Ye, K.V. Breugel, Simulation of connectivity of capillary porosity in hardening cement-based systems made of blended materials, HERON 54 (2/3), 2009, pp. 163–184.

[57] American Concrete Institute, ACI 522R-10 Report on Pervious Concrete. Farmington Hills, USA, 2010.

[58] A.S. Agar-Ozbek, et al., Investigating porous concrete with improved strength: Testing at different scales, Const. Bldg. Mat. 41 (2013) 480–490.

[59] D. Breton, Contribution to the formation mechanism of the transition zone between rock-cement paste, Cement Concr. Res. 23 (2) (1993) 335–346.

[60] J.L. Clarke, Structural Lightweight Aggregate Concrete, Blackie Academic & Professional, London, 1993.

[61] C.G. Rocco, M. Elices, Effect of aggregate shape on the mechanical properties of a simple concrete, Eng. Fract. Mech. 76 (2) (2009) 286–298.

[62] T. Sofilić, et al., Radionuclides in metallurgical products and waste, Acta Metall. Slov. 10 (1) (2004) 29–35.

[63] L.A. Wendling, et al., Geochemical and ecotoxicological assessment of iron- and steel-making slags for potential use in environmental applications, Environ. Toxicol. Chem. 32 (11) (2013) 2602–2610.

[64] European Committee for Standardization, EN 12457-2:2002: Characterisation of waste. Leaching. Compliance test for leaching of granular waste materials and sludges. One-stage batch test at a liquid to solid ratio of 10 l/kg for materials with particle size below 4 mm (without or with size reduction), CEN/TC 292, Brussels, 2002.
[65] I. Barišić, I. Netinger Grubeša, B.H. Kutuzović, Multidisciplinary approach to the environmental imapct of steel slag reused in road construction, Road Mater Pavement (2016) accepted for publication.
[66] European Commission, Study on methodological aspects regarding limit values for pollutants in aggregates in the context of the possible development of end-of-waste criteria under the EU Waste Framework Directive, Technical Report by the Joint Research Centre of the European Commission, Luxembourg, Luxembourg, 2014.
[67] International Standardization Organization, ISO 11466, Soil quality—Extraction of trace elements soluble in aqua regia, ISO/TC 190, 1995, Geneva, Switzerland, 1995.
[68] International Standardization Organization, ISO 11047, Soil quality—Determination of cadmium, chromium, cobalt, copper, lead, manganese, nickel, and zinc in aqua regia extract of soil—Flame and electrothermal atomic absorption spectrometric methods, ISO/TC 190, 1998, Geneva, Switzerland, 1998.
[69] European Commission, Directive 2000/60/EC of the European Parliament and of the Council establishing a framework for the Community action in the field of water policy, Official Journal of the European Union L 372, European Parliament and Council of the European Union, Brussels, Belgium, 2000.
[70] European Commission, Directive 2006/118/EC of the European Parliament and of the Council of 12 December 2006 on the protection of groundwater against pollution and deterioration (Daughter to 2000/60/EC), Official Journal of the European Union L 372, European Parliament and Council of the European Union, Brussels, Belgium, 2006.
[71] European Commission, Directive 2006/11/EC of the European Parliament and of the Council of 15 February 2006 on pollution caused by certain dangerous substances discharged into the aquatic environment of the community, Official Journal of the European Union L 372, European Parliament and Council of the European Union, Brussels, Belgium, 2006.
[72] P. Chaurand, et al., Environmental impacts of steel slag reused in road construction: A crystallographic and molecular (XANES) approach, J. Haz. Mat. 139 (3) (2007) 537–542.
[73] Ministry of Agriculture, Croatian Government, Ordinance on emission limit values for wastewater discharges, 2010, Zagreb, Croatia: Official Gazette 87/10.
[74] E.E.V. Chapman, et al., A review of metal (Pb and Zn) sensitive and pH tolerant bioassay organisms for risk screening of metal-contaminated acidic soils, Environ. Poll. 179 (2013) 326–342.
[75] J.L. Camas-Anzueto, et al., Sensitive layer based on lophine and calcium hydroxide for detection of dissolved oxygen in water, Meas 68 (2015) 280–285.
[76] J. Colt, Water quality requirements for reuse systems, Aqua. Eng. 34 (2006) 143–156.
[77] N.M. Piatak, M.B. Parsons II RRS, Characteristics and environmental aspects of slag: A review, Appl. Geochem. 57 (2014) 236–266.
[78] W.M. Mayes, P.L. Younger, J. Aumonier, Buffering of alkaline steel slag leachate across a natural wetland, Environ. Sci. Tech. 40 (2006) 1237–1243.
[79] M. Sakata, Movement and neutralization of alkaline leachate at coal ash disposal sites, Environ. Sci. Tech. 21 (1987) 771–777.
[80] Organisation for Economic Co-operation and Development (OECD), OECD Guidelines for the Testing of Chemicals, Section 2: Effects on Biotic Systems, 1984, Test No 207: Earthworm, Acute Toxicity Tests.

The Indian experience of steel slag application in civil engineering

7

7.1 Introduction

India is the third-highest producer of crude steel in Asia, according to the World Steel Association (WSA) report 2014 [1]. The majority of steel slag is used for earth filling in low laying areas. Slag chemical properties are very similar to those of natural aggregates, and steel slag can be used as a replacement for them. In addition, steel slag can be used as aggregates in filters in wastewater engineering; it also positively affects the removal of various heavy metals from wastewater.

There are many steelmaking plants in India. The production of iron or steel is associated with the generation of solid waste material like slag. Usually, 1 metric ton of steel production produces 1 tonne of solid waste. Large steelmaking plants generates about 29 million metric ton of waste material annually [2].

Slag is an ideal material to use in the base course and subbase course. It can transfer heavy loads very easily and also prevents water from penetrating the subgrade soil, which will help in reducing heaving due to expansion of the soil beneath the surface. Slag aggregates also contribute to durability and economy because it is more cost-effective than the natural aggregates they replace [3].

Steel slag reduces the porosity and permeability of soil, which increases waterlogging. It can be used as pavement material in many different forms. It can be used as a base or subbase material, either in a waterbound or unbound condition. It meets all the requirements of the Ministry of Road Transport and Highways (MoRTH). As per the Indian Road Congress (IRC) IRC:37–2001, Rs 5 lakh/km can be saved by using slag when constructing flexible road when it is used from the bottom subbase layer to the top [2].

Over the last 15 years, test roads have been constructed on an experimental basis in Visakhapatnam (AP), Bokaro (Bihar), Rourkela (Orissa), and Delhi using waste material from steel plants on an experimental basis [4].

This chapter describes an experiment performed to evaluate the strength characteristics of steel slag. The objective of this study was to reduce the use of natural resources and promote the use of waste materials. The physical properties, chemical properties, and engineering properties of this material, such as sieve analysis of fine and coarse aggregate and specific gravity, were studied. Slag was evaluated for strength by measuring the direct shear strength, using the California bearing ratio (CBR).

Characteristics and Uses of Steel Slag in Building Construction. http://dx.doi.org/10.1016/B978-0-08-100368-8.00007-5

7.2 Analysis of Indian steel slag

7.2.1 Physical properties of steel slag

The physical characteristics of slag, like weight, size of particles, and structural properties, vary according to the type. Processed steel slag is strong, hard, durable, and dense, and consists of roughly cubical particles that make it especially suitable for use in road construction. Steel slag is not expansive in nature, but if so water quenching is the first step in the weathering of steel slag, which may be essential to provide a stable and non-expansive construction material.

In the cooling process, lime in the slag may be hard burned. It has a hard outer shell or surface that masks the unsound, soft inner core of unhydrated lime. Lime must be saturated by water in order for the particle to be stable. In Table 7.1, the physical properties of Indian steel slag are shown.

7.2.2 Chemical properties of steel slag

Table 7.2 shows the mineralogical and chemical composition of steel slag. It can be seen that the iron percentage in the steel slag is very less.

Table 7.1 **Physical properties of Indian steel slag**

Physical Property	Steel Slag
Colour	Black
Specific gravity	2.76
Appearance	Crystalline
Compacted unit weight (kN/m^3)	20.98–23.34
Absorption (%)	0.93% of total weight

Table 7.2 **Chemical composition of steel slag**

Element	Weight (%)
$CaCO_3$	9.65
SiO_2	51.82
MgO	1.13
Al_2O_3	1.78
Ca	17.98
Ti	0.38
Cr	0.39
Mn	2.14
Fe	6.44

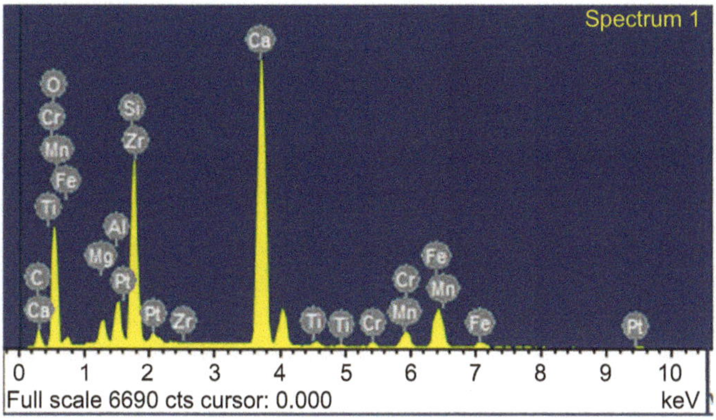

Figure 7.1 Energy dispersive spectroscope analysis of steel slag

Also, calcium oxide and calcium are at lower levels, so expansion will be less of a problem. Here, slag can be used as an aggregate in the construction of roads. Fig. 7.1 represents the energy dispersive spectroscope analysis of steel slag.

7.3 Experimental analysis

In India, Maharashtra Industrial Development Corporation (MIDC) Jalna was declared as "steel industrial zone of Maharashtra State" by the Government of Maharashtra State. Metaroll (Rajuri Steel Industries), Mauli Steel Private Limited, S.R.J Pitty Steel Pvt. Ltd., Kalika steel are some of the steel manufacturers of the MIDC Jalna; steel slag is collected from these manufacturers. After the collection of steel slag, the geotechnical properties and particle size distribution curve are determined. The values of liquid limit, plastic limit, and plasticity index are determined by Atterberg's limit apparatus, and the values of maximum dry density (MDD) and optimum moisture content (OMC) are determined by the Standard Proctor test apparatus. Using the CBR test apparatus, the California bearing ratio (CBR) value is determined. To determine the CBR value, the sample is soaked for 96 h, and dry steel slag was taken.

7.3.1 Engineering properties of slag

7.3.1.1 Determination of the specific gravity of steel slag

Specific gravity determines the heaviness or lightness of a material with reference to a standard material. Slag particles with more specific gravity are stabler and stronger. However, this does not indicate the behavior of a soil mass under external loads. Specific gravity is also useful for soil mineral classification and hydrometer analysis. In addition, it is useful in the computation of a hydraulic gradient. For soils containing soluble salts, kerosene may be preferred instead of distilled water. Conventionally oven-dried soil is used [5].

From these details, the specific gravity of the slag is 2.71, which is greater than 2.65. This shows that the steel slag is strong and may not heave. (The density of boulder filling should not be less than 2.65 be as stated in MoRTH.)

7.3.1.2 Water content and water absorption of steel slag

The water content of slag is found by oven drying. In this method, slag samples are oven dried at 105°C to 110°C for 24 h and found the difference between the weights before putting and after removing it from oven and water content of the steel slag found to be 7.62%. For water absorption testing, these aggregates are placed in the water for 24 hours and again the difference of weights is taken before placing in water and after withdrawing from water. The water absorption of steel slag has been determined and is found to be 0.93%.

7.3.1.3 Grain size distribution of steel slag

Steel slag consists of different particle sizes in various quantities (Table 7.3). Particle size and their distribution affect engineering properties. Sieves are used for separating the different sizes. Sieving can easily separate the coarser and finer particles. Air-drying is enough if the soil contains less than 5% moisture content, and dry analysis may be adopted. Coarse-grained soils with little or no fine particles then it needs dry analysis only (as per MoRTH, Grading requirements of coarse aggregates are shown in Table 7.5. Coarse aggregates must be in the range as given in table 7.5). If the soil contains more than 5% fines particles (passing 75 μ fraction) then these samples should be dipped in water for 24 h. Wet analysis is necessary to separate particles smaller than 75 μ. The cumulative percentage by weight retained on any sieve is the sum of the percentages by weight retained on all the sieves up to and above that sieve.

Percentage passing on any sieve (Percentage finer) =
100 – Cumulative percentages by weight retained by sieve

From Fig. 7.2, steel slag aggregates are well graded and can be used in construction of waterbound macadam road [6].

Table 7.3 Grain size distribution of steel slag

Sr. No.	IS Sieve (mm)	Mass of Slag Retained (g)	Cumulative Mass Slag Retained (g)	Cumulative Percentage Retained	Percentage Finer
1	75	0	0	0	100
2	53	1522	1522	15.22	84.78
3	26.5	3312	4834	48.34	51.66
4	9.5	1852	6686	66.86	33.14
5	4.75	1429	8115	81.15	18.85
6	2.36	685	8800	88	12
7	0.425	443	9243	92.43	7.57
8	0.075	452	9695	96.95	3.05
9	Pan	305	10000	100	0

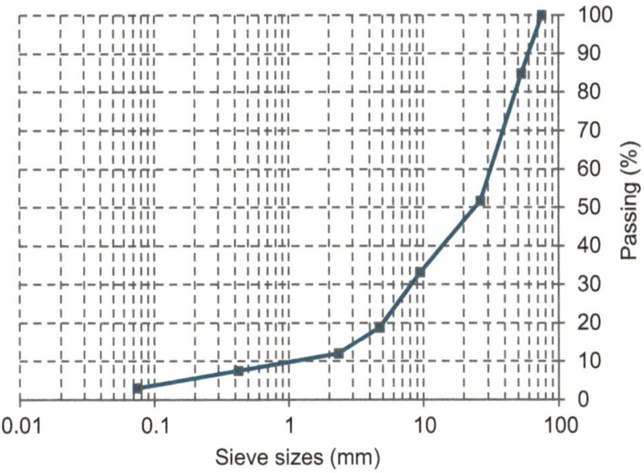

Figure 7.2 Particle size distribution of steel slag.

Table 7.4 **Grading for close-grained subbase materials**

IS Sieve Designation	Percentage by Weight Passing the IS Sieve		
	Grading I	**Grading II**	**Grading III**
75.00 mm	100	–	–
53.00 mm	80–100	100	–
26.50 mm	55–90	70–100	100
9.50 mm	35–65	50–80	65–95
4.75 mm	25–55	40–65	50–80
2.36 mm	20–40	30–50	40–65
0.425 mm	10–25	15–25	20–35
0.075 mm	3–10	3–10	3–10

Adopted from Ministry of Road Transport and Highways. Specification of road and bridge works, Section 300 and 400, 2000.

The material to be used for the work shall be natural sand, murum, gravel, crushed stone, or a combination thereof, depending upon the grading required. Materials like crushed slag, crushed concrete, brick metal, and kankar may be allowed only with the specific approval of the project manager. The material shall be free from organic or other deleterious constituents and conform to one of the three gradings given in Table 7.4.

Slag passes through 425 μ (0.425 mm) sieve for all the grading. According to IS: 2720 (Part 5), liquid limit and plasticity index should not be more than 25% and 6%, respectively [7].

Depending upon the grain size distribution, one can decide whether soil should be used. As stated by the specification manual of MoRTH, the grading of the material is very close to the type-II aggregates used for the specification. All the other tests are required before a decision can be made on using the aggregates.

Table 7.5 **Grading requirements of coarse aggregates as per MoRTH**

Grading Size Range	IS Sieve Designation	Percentage of Aggregates Passing by Weight
1. 90–45 mm	125 mm	100
	90 mm	90–100
	63 mm	25–60
	45 mm	0–15
	22.4 mm	0–5
2. 63–45 mm	90 mm	100
	63 mm	0–100
	53 mm	25–75
	45 mm	0–15
	22.4 mm	0–5
3. 53–22.4 mm	63 mm	100
	53 mm	95–100
	45 mm	65–90
	22.4 mm	0–10
	11.2 mm	0–5

Adopted from Ministry of Road Transport and Highways. Specification of road and bridge works, Section 300 and 400, 2000.

7.3.1.4 Determination of Atterberg's limits

Atterberg's limits are liquid limit, plastic limit, and plasticity index. The liquid limit is the moisture content at which the groove, formed by a standard tool into the sample of soil taken in the standard cup, closes for 10 mm on being given 25 blows in a standard manner. At this limit the soil possess low shear strength. The identification and classification of fine-grained soil is done with the help of liquid and plastic limits. For cohesion-less soil, however, the liquid limit is meaningless. The potential for swelling can be indicative from high liquid limits coupled with a high plasticity index. Soil with low clay content has to be tested soon after being mixed thoroughly with water. In the case of tearing of the sides of groove in Attergberg's limit apparatus, the groove may be cut in stages. Soils sometimes tend to slide instead of flowing. In such cases, discard the result and repeat the process until flowing takes place. If there is continued sliding, report that the liquid limit could not be obtained. Most clay has a liquid limit of the order of 50%–90%; however it can exceed 100%.

Consistency in fine-grained soils means the properties of friability, stickiness, plasticity, and resistance to deformation. They can be categorized as very hard, hard, stiff, medium, soft, and very soft. The difference between liquid and plastic limits is called the plasticity index, and it represents the range of water content over which the soil is plastic. The plasticity index, in combination with water content at liquid limit (wL), indicates how sensitive the soil is to the changes in moisture content. Sands, gravel, and peat do not posses plasticity and are reported as being nonplastic (wP = 0). Both liquid and plastic limits depend upon the amount and type of clay present in soil (IS-2720-Part 5-1995).

The liquid limit of steel slag cannot be obtained due to noncohesiveness or reduced cohesiveness of the particles. The plastic limit and the plasticity index of steel slag cannot be estimated due to the fact that steel slag is low-plastic in nature.

7.3.1.5 Shrinkage limit determination

From these observations, the average value of the shrinkage limit is 12.90, and volumetric shrinkage is 0.66%. At the shrinkage limit, if water content is reduced further, air enters the voids of the soil and keeps the volume of voids constant. From the shrinkage limit, shrinkage factors like shrinkage ratio, shrinkage index and volumetric shrinkage can be determined. This limit is needed for studying the swelling and shrinkage properties of cohesive soil. The shrinkage factor helps in the design problems of structure made up of this soil or resting on such soil. It helps in assessing the suitability of soil as a construction material in foundations, roads, embankments, and dams. The state of the given soil can also be found from the shrinkage limit. Volumetric shrinkage is the decrease in volume (expressed as a percentage of the soil mass when dried) of a soil mass when the water content is reduced from a given percentage to the shrinkage limit. The shrinkage index is the numerical difference between the plastic limit and shrinkage limit of the remolded soil [8].

7.3.1.6 Maximum dry density and optimum moisture content for steel slag

Density is the most easily determined property of soil. It is the basic parameter to which all other engineering properties are related. The density of soil in compacted soil is used as the quality control measure of earthwork in embankments. In natural soil, it helps to determine bearing capacity, calculation of settlement, and the stability analysis for slopes. Bulk unit weight varies with the type of soil, moisture content, and compaction. Higher unit weight means higher shear strength, greater imperviousness (low permeability), less compressibility, less settlement, and greater ultimate bearing capacity. Engineering properties of soil are improved with increases in the unit weight of soil from Fig.7.3 and Table 7.6, it is clear that OMC is 9.62% and MDD is 2.39 gm/cc [9].

7.3.1.7 Determination of California Bearing Ratio

CBR is defined as the ratio of the test load to the standard load, expressed as the percentage for 2.5 and 5.0 mm penetration of the plunger in soil sample. It is considered as to be one of the most commonly used and widely accepted tests. It can be used to analyse existing pavements, layers by layer, with respect to their strength and load-carrying capacity. The CBR test also helps in identifying the cause of road pavement failure.

The CBR values are usually calculated for the penetration values 2.5 mm and 5.0 mm. Generally, the CBR values at 2.5-mm penetration will be greater than that at 5-mm penetration. In that case, the former is to be taken as the CBR values for design purposes.

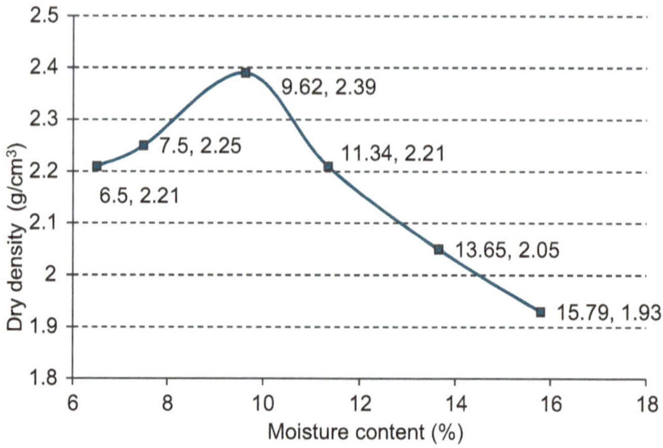

Figure 7.3 Water contents versus dry density (9.62% optimum moisture content, 2.39g/cm3 is the maximum dry density).

If the CBR values corresponding to a penetration of 5 mm exceed that for 2.5 mm, the test is repeated. If identical results are obtained again, the CBR value corresponding to 5-mm penetration is taken for design purposes. The CBR value shall be reported to the first decimal place [9].

To determine the CBR test for unsoaked steel slag, do the following:

Observations:

Optimum moisture content = 9.62%

Weight of empty mould = 3662 gm

Weight of mould + compacted soil = 8987 gm

Weight of compacted soil = 5325 gm

Volume of mould = 2032.21 cm^3

Bulk density = 2.62 gm/cc

Dry density = 2.39 gm/cc

CBR = (Pt/Ps) × 100

where Pt = the corrected unit test load and Ps = the unit standard load for the same depth of penetration.

CBR (2.5 mm) = (350.98/1370) × 100 = 25.62%

CBR (5.0 mm) = (420.74/2055) × 100 = 20.43%

CBR value of unsoaked steel slag is 25.62%

To determine the CBR for a wet steel slag sample, do the following:

Observations:

Optimum moisture content = 9.62%

Weight of empty mould = 3662 gm

Weight of mould + compacted soil = 8987 gm

Weight of compacted soil = 5325 gm

Volume of mould = 2032.21 cm^3

Bulk density = 2.62 gm/cc

Dry density = 2.39 gm/cc

Table 7.6 **Determination of maximum dry density and optimum moisture content for steel slag**

	Determination No.	**1**	**2**	**3**	**4**	**5**	**6**
Sr. No.	**Percentage of Water Added**	**4%**	**6%**	**8%**	**10%**	**12%**	**14%**
1	Mass of the empty mould (wm) (gm)	3510	3510	3510	3510	3510	3510
2	Mass of mould + comp. soil (w) (gm)	5865	5935	6129	5969	5831	5740
3	Mass of compacted soil (gm)	2355	2425	2629	2459	2321	2230
4	Volume of mould (Vm) (ml)	999.81	999.81	999.81	999.81	999.81	999.81
5	Mass of dry soil (gm)	2000	2000	2000	2000	2000	2000
6	Bulk density (g/ml) $\gamma = (w - wm)/Vm$	2.35	2.42	2.62	2.45	2.32	2.23
7	Moisture content (w) (%)	6.50	7.50	9.62	11.34	13.65	15.79
8	Dry density (g/ml) $\gamma d = \gamma/(1 + w)$ (g/cm^3)	2.21	2.25	2.39	2.21	2.05	1.93

From the proving ring graph that results, by interpolation, we get

CBR (2.5 mm) = 292.12 /1370 × 100 = 21.32

CBR (5 mm) = 390.22 /2055 × 100 = 18.99

The CBR value of unsoaked steel slag is 21.32%. The detailed observations are expressed in Table 7.7 and results are represented in Fig. 7.4.

The determination of abrasion value of steel slag

Observations:

The aggregate abrasion value should not exceed 30% by weight of the aggregates for the wearing surface. In addition, it should not increase 50% by weight of the aggregate for concrete to be used for nonwearing surface [IS: 2386 (Part 4)]. The aggregate abrasion value can be determined as:

Weight of original slag sample (w1) = 10 kg

Weight of coarser material than 1.7-mm IS sieve (w2) = 7 kg

Percentage of wear:

$$\textbf{percentage wear} = 100\times\frac{(\mathbf{w1}-\mathbf{w2})}{\mathbf{w1}}$$

$$= 100 \times \frac{(10-7)}{10}$$

$$= 30\%$$

The abrasion value of slag is within IS specifications; therefore, this aggregate can be used for the road as the abrasion is within specified limits.

The determination of steel slag impact value can be done as follows:

The aggregate impact value should not exceed 45% by weight of the aggregate for concrete other than the wearing surface. The aggregate impact value should not exceed 30% by weight for a concrete wearing surface, such as runways, roads, pavements, and floors [10], IS: 2386 (Part 4) Average impact value of steel slag is 13.90%

The slag has a good level of strength for impact loading, as the value is within the IS specified limits.

The determination of crushing value of slag is as follows:

The aggregate impact value should not exceed 45% by weight of the aggregate for concrete other than wearing surface. The aggregate impact value should not exceed 35% by weight for concrete wearing surface such as runways, roads, pavements, floors etc. [10] IS: 2386 (Part 4). The average crushing value of the steel slag aggregate is 5.20%

The compacted thickness for a Grade 1 layer shall be 100 mm, while for Grade 2 and Grade 3 layers, it shall be 75 mm. The binding material to be used for water-bound macadam as a filler material meant to prevent raveling must be designated as suitable by the engineer with a plasticity index (PI) value of less than 6, in accordance with IS: 2720 (Part 5) [7].

Table 7.7 **California bearing ratio test observations**

Penetration (mm)	Dial Gauge Reading	Load	CBR (%)	Dial Gauge Reading	Load	CBR (%)
	Unsoaked			Soaked		
0	0	0		0	0	
0.5	66	140.74		64	138.02	
1	98	214.14		94	204.68	
1.5	134	283.46		112	244.56	
2	156	327		129	279.64	
2.5	161	350.98	25.62	134	292.12	21.32
3	173	371.12		144	306.02	
3.5	178	388.04		146	317.98	
4	180	392.4		158	344.44	
4.5	188	409.84		167	364.06	
5	193	420.74	20.47	179	390.22	18.99
5.5	210	457.8		188	409.84	
6	213	462.16		193	418.56	
6.5	218	475.34		201	438.58	

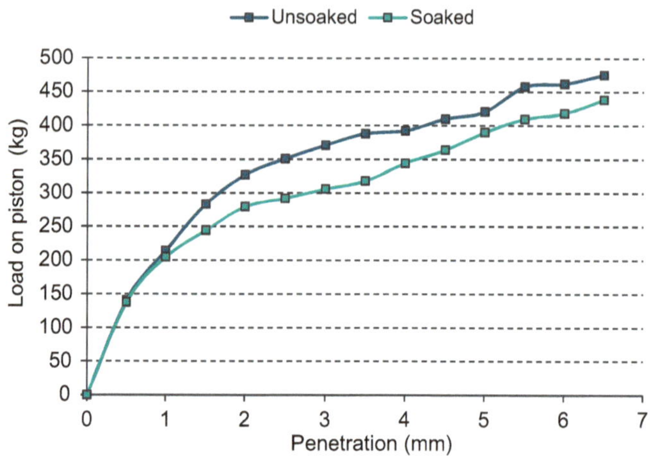

Figure 7.4 Load against a penetration curve

The quantity of binding material to be used depends on the type of screenings. Generally, the quantity required for 75-mm compacted thickness of water-bound macadam will be 0.06–0.09 $m^3/10\ m^2$ and 0.08–0. 10 $m^3/10\ m^2$ for 100-mm compacted thickness. The abovementioned quantities should be taken only as a guide for estimating quantities for construction and other purposes. Application of binding materials may not be necessary when the screenings used are of crushable type such as murum or gravel.

The surface of the subgrade/subbase/base to receive the water-bound macadam course shall be prepared to the specified guidelines provided by MoRTH and cross fall (camber) arid (which is provided for road to remove rainwater quickly from road surface) made free of dust and other extraneous material. Any ruts or soft yielding spots shall be corrected in an approved manner and rolled until a firm surface is obtained if necessary by sprinkling water. Any subbase/base/surface irregularities, where predominant, shall be made good by providing the appropriate type of profile corrective course (leveling course) to Clause 501 of these specifications. As far as possible, laying water-bound macadam course over an existing thick bituminous layer may be avoided since it will cause problems with internal drainage of the pavement at the interface of two courses. It is desirable to completely pick out the existing thin bituminous wearing course where water-bound macadam is supposed to be laid over it. However, where the intensity of rain is low and the interface drainage facility is efficient, water-bound macadam can be laid over the existing thin bituminous surface by cutting 50-mm × 50-mm furrows at an angle of 45 degrees to the center line of the pavement at 1-m intervals into the existing road. The directions and depth of the furrows shall be such that they provide adequate bondage and serve to drain water to the existing granular base course beneath the existing thin bituminous surface.

The physical requirements of coarse aggregate as per MoRTH are listed in Table 7.8.

Table 7.8 **Physical requirements of coarse aggregate**

Sr. No.	Test	Test Method	Requirement
1	Los Angeles abrasion value Aggregate impact value	IS: 2386 (Part 4) [11]	40% maximum
2	Combined flakiness and elongation indices (total)	IS: 2386 (Part 4) [11] or IS: 5640 [13]	30% maximum
3	Water absorption	IS: 2386 (Part 3) [12]	2% maximum
4	Liquid limit of material passing 425 μ	IS: 2720 (Part 5) [7]	Not more than 25%
5	Plasticity index of material passing 425 μ	IS: 2720 (Part 5) [7]	Not more than 6%

Aggregate may satisfy the requirements of either of two tests
If the water absorption is more than 2%, a soundness test shall be carried out as per IS: 2386 (Part 5)
Adopted from Ministry of Road Transport and Highways. Specification of road and bridge works, Section 300 and 400, 2000.

7.3.2 Testing of natural aggregates

7.3.2.1 Sieve analysis

In mechanical soil stabilization, the main principle is to mix a few selected soils in such a proportion that a desired grain size distribution is obtained for the design mix. Hence for proportioning the selected soils, the grain size distribution of each soil is to be first known. From sieve analysis, it can be concluded whether aggregates are well-graded gravels, poorly graded gravels, well-graded sand, or poorly graded sand. Fig. 7.5 shows that aggregates are well-graded gravels.

Table 7.9 gives the sieve analysis for natural aggregate.

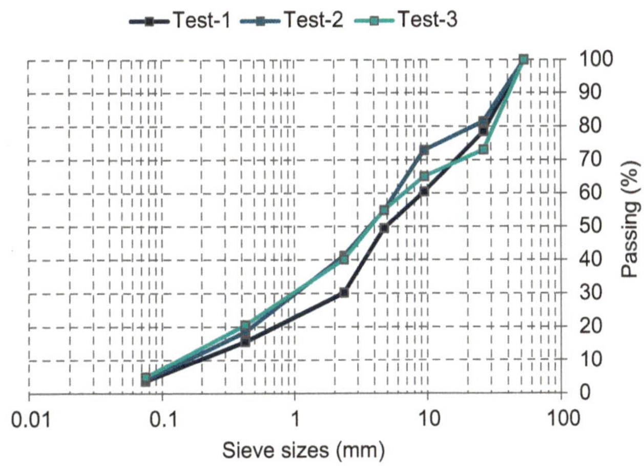

Figure 7.5 Sieve analysis of natural aggregate

Table 7.9 Sieve analysis for natural aggregate

Sr. No.	IS Sieve Size	Test 1	Test 2	Test 3
1	53	100	100	100
2	26.5	78.5	81.45	73.05
3	9.5	60.5	72.85	65.00
4	4.75	49.5	54.95	55.00
5	2.36	30.25	41.20	40.15
6	0.425	15.5	18.50	20.54
7	0.075	3.6	4.05	4.95

Table 7.10 Determination of maximum dry density (MDD) and optimum moisture content (OMC) for aggregates

Sr. No.		Test 1	Test 2	Test 3
1	Determination of OMC (%)	13.10	13.15	13.11
2	Determination of MDD (gm/cc)	1.85	1.86	1.84

Table 7.11 Determination of CBR for aggregates

Sr. No.		Test 1	Test 2	Test 3
1	Determination of OMC (%)	13.15	13.14	13.13
2	Determination of MDD (gm/cc)	1.84	1.85	1.86
3	Determination of CBR (%)	21.58	23.08	22.13

Typical Atterberg's limits of the natural aggregates

Average liquid limit = 23.23%

Average plastic limit = 17.79%

Average plasticity index = 5.31%

Table 7.10 gives the determination of MDD and OMC for aggregates.

The results shown in Table 7.10 are at average, 13.12% for OMC and 1.84 gm/cc for MDD which has been represented in Fig. 7.2.

Table 7.11 shows the determination of the CBR value for aggregates.

The results obtained from Table 7.11 are at average 13.14% for OMC, 1.85 gm/cc, for MDD, and 22.26% for CBR.

The determination of impact value, flakiness index, and water absorption for the aggregate are shown in Table 7.12.

The average of the impact value is 11.89%, flakiness index is 22.42%, and water absorption is 1.19%.

Table 7.12 **Impact value, flakiness index, water absorption for the aggregate**

Sr. No.	Determination of Values (%)	Test 1	Test 2	Test 3
1	Impact value	11.55	12.05	12.09
2	Flakiness index	22.35	22.51	22.40
3	Water absorption	1.50	1.05	1.04

Table 7.13 **Comparison of steel slag aggregate versus natural aggregates**

No.	Test		Steel Slag Aggregate	Natural Aggregate
1	Water content		9.67%	14.28%
2	Dry density		2.39 gm/cm^3	1.81 gm/cm^3
3	Specific gravity		2.71	2.60
4	Plasticity index, plastic limit, and liquid limit		Nonplastic	4.5% 20.5%
5	MDD		2.39 gm/cm^3	1.85 gm/cm^3
6	OMC		9.62%	13.14%
7	CBR	Unsoaked at 2.5 mm	25.62%	22.26%
		Soaked at 2.5 mm	21.32%	17.6%
8	Impact value		13.90%	11.89%
9	Flakiness index		4.80%	22.42%
10	Water absorption		1.0%	1.19%
11	Aggregate crushing value		5.20%	10.21%

7.4 Comparative analysis of steel slag aggregates and soil aggregates

From Table 7.13, comparison of steel slag aggregates versus natural aggregates, it is possible to use steel slag as aggregates in the construction of road. As main important tests such as, abrasion, water absorption, crushing, impact, MDD and OMC also show positive results.

7.5 Cost analysis

Cost analysis for road construction depending upon finance required for the various construction activities of road construction by utilizing steel slag.

The abstract information for a 1-km length of road is given in Table 7.14 (lead and lift is considered for only 5 km). For construction of road there will be various stages such as cleaning of site, excavation for WBM road, rolling (compaction) at various stages, supply and laying of materials, etc.

Table 7.14 **Abstract sheet for 1-km road**

Sr. No.	Quantity (m^3)	Description	Unit	Rate	Amount
1	112.5	Supplying 60-mm trap/granite/quartzite/gneiss stone oversize metal at the roadside, including conveying and stacking etc. (including blasting)	Per cubic meter	382.56	Rs 43,038
2	375	Supplying 40-mm trap/granite/quartzite/gneiss stone including conveying and stacking etc. (including blasting)	Per cubic meter	397.67	Rs 1,49,126
3	237.51	Supplying hard murum/kankar at the roadside, including conveying and stacking	Per cubic meter	195.64	Rs 52,336
4	93.57	Supplying soft murum at the roadside, including conveying and stacking	Per cubic meter	230.56	Rs 21,573

As like spreading of materials, compacting of materials, and excavation these activities will be the same as use of the steel slag. Material only metal like, 60 mm aggregates, 40 mm aggregates hard murum and the soft murum that will be replaceable as the properties of the steel slag aggregates and the natural aggregates that will be replaced. The total cost required for the road construction is Rs 266,073.00. For the hundred percentage replacement of steel slag, it requires only the transportation charges. It means that Rs 199,555.00 can be saved by replacing natural aggregates with steel slag aggregates (considering 25% transportation charges).

7.6 Comparative analysis between MoRTH standard and experimental method

From Table 7.15, specific gravity, crushing value, abrasion value, water absorption impact values show positive results, and as slag is nonplastic and very less incohesive, steel slag can be used as aggregates in road construction (cannot be obtained because of less cohesion and low plastic in nature).

7.7 Conclusion

In India, systems are developed for utilization of steel slag and to raise awareness among industrialists and the public that how slag can be used in construction. Figs. 7.6–7.8 are some examples where steel slag is used for construction of water bound macadam road. It is very important that steel slag that is used in construction be obtained from a supplier who has a quality control program in place to minimizes the amount of unsound particles. From the analysis given in this chapter, the characteristic strength of steel slag aggregates and natural soil aggregates are about equal, and some results are found to be more than equal. It means that the steel slag is found to be stronger as like natural soil aggregates. The conclusions to be drawn here are as follows:

- The abrasion value of steel slag aggregates found to be 25.66% which is within the specified limits as per MoRTH, and IS: 2386 (Part 4). It should be less than 30%.
- With regard to impact value, steel slag is suitable for use instead of natural aggregates. The impact value of the steel slag aggregates is found to be 13.90% of the total weight as per IS: 2386 (Part 4) and MoRTH. It should be less than 30%. So these steel slag aggregates can be used to replace natural aggregates.
- The crushing value of the steel slag aggregate is found to be 5.20%.
- The CBR value of the steel slag aggregates is found to be 21.32% and 25.62% for soaked and unsoaked conditions, respectively.
- The impact value of steel slag aggregates is found to be 13.90%. The aggregate impact value should not exceed 45% by weight of the aggregate for concrete other than the wearing surface. The aggregate impact value should not exceed 30% by weight for concrete wearing surfaces such as runways, roads, pavements, and floors [IS: 383-1970, IS: 2386 (Part 4)].
- Due to the rough surface of steel slag particles, they prevent the skidding of vehicles. Due to this effect, on sloping and curved roads, slag can be used as skidding resistance particles on the topmost layer.

Table 7.15 Comparison between MoRTH and experimental analysis

Sr. No.	Property of Steel Slag	Analytical Method (as per MoRTH) Natural Aggregates	Experimental Method (results of steel slag)
1	Specific gravity	Minimum 2.6	2.71
2	Grain size distribution	Well-graded	Well-graded
3	Bulk density	1.52 to 1.75 gm/cm^3	2.62 gm/cm^3
4	Dry density	N/A	2.39 gm/cm^3
5	Liquid limit	Not more than 25%	Cannot obtain
6	Plastic limit	N/A	Cannot obtain
7	Plasticity index	Not more than 6%	Cannot obtain
8	Shrinkage limit	N/A	12.90%
9	MDD	N/A	2.39 gm/cm^3
10	OMC	N/A	9.67%
11	CBR	20–30%	21.32–25.62%
12	Aggregate crushing value	Less than 15%	5.20%
13	Aggregate abrasion value	30–40%	30%
14	Flakiness and elongation index	Maximum of 40%	4.80%
15	Water absorption	Maximum of 2%	0.93%
16	Aggregate impact value	Maximum of 30%	13.90%

Figure 7.6 Approach road for Aurangabad–Jalna Highway MS, India

Figure 7.7 WBM approach road Aurangabad–Jalna, MS, India

Figure 7.8 Approach road to the Dhanlaxmi Steel Ind. Pvt. Ltd at MIDC, Jalna, MS, India

7.8 Scope of future research

If steel slag is used, natural resources can be preserved in steel industrial areas. Slag can be used for various purposes. There is much more to explore about steel slag as a civil engineering material, including the following:

- The study of friction created between the road and vehicle tyres
- The effect of temperature on steel slag aggregates
- The effect of steel slag and plastic waste as a binder on the properties of soil
- The study of the properties of steel slag when used with hot mix binder
- The use of steel slag as aggregate in concrete

References

[1] World steel production report on, "Crude steel production", September 2014.
[2] J.K. Mohapatra, B.P. Chandrasekhar, Rural Roads, Indian Infrastructure Report, 2007, pp. 109–137.
[3] D. Lewis, Construction of air-cooled blast furnace slag base courses, Journal of National Steel Slag Association (2003) 1–5, MF 183–184.
[4] P. Kumar, A. Kumar, Steel industry waste utilization in the road sector of India, J. Inst. Eng., India 80 (2000) 182–185.
[5] Indian Standard Code. IS: 2720, Part 3, Section 1, 2 and 3, 1980 and reaffirmed in 1987, for determination of specific gravity.
[6] Indian Standard Code. IS: 2720, Part 4, 1995 for grain size analysis.
[7] Indian Standard Code. IS: 2720, Part 5, 1995 for determination of liquid limit and plastic limit.
[8] Indian Standard Code. IS: 2720, Part 6, 1972; reaffirmed in 1978, for determination of shrinkage factor.
[9] Indian Standard Code. IS: 2720, Part 16 for laboratory determination of CBR.
[10] Indian Standard Code. IS: 383–1970 for determination of quality of aggregates.
[11] IS 2386 Part 4, 1963 reaffirmed 1990, 1997 Determination of mechanical properties of aggregates.
[12] IS 2386 Part 3, 1963 reaffirmed 1990, 1997 Determination specific gravity, density, voids, absorption and bulking of aggregates.
[13] IS 5640, 1970, reaffirmed 1998, 2003, Determination of aggregate impact value.

Recommendations for future research

8

Because more and more structures are built every year, there is a growing demand for raw materials and energy. Until recently, natural materials were used predominatly in civil engineering projects, but the amount of these is limited. In modern civil engineering, the use of unconventional materials is becoming more and more frequent. Figure 8.1 illustrates current and future trends in civil engineering practices, which in some way were the impetus for the authors to write this book.

Some materials that are being increasingly used in this area are fly ash, glass, used tiers, and various kinds of slag. Use of these materials not only reduces the space necessary for their disposal, since they are waste materials/by-products, but also the energy necessary for obtaining materials that are usually used in civil engineering. In addition, their use allows reserves of natural materials like gravel and sand, which until now had been essential for civil engineering, to be preserved.

A large quantity of stone, gravel, or sand is used daily in civil engineering. This devastation of natural wealth and exhaustion of natural stocks can be prevented by the use of unconventional materials like waste materials and industrial by-products. The use of those materials contributes to a more rational use of good-quality aggregate and also helps resolve environmental problems that arise from the disposal of waste material. With new structures being built around the world every day, millions of tons of raw materials are used and natural stocks are impoverished. At the same time, developed countries must confront a growing problem with waste materials. However, small countries such as Croatia share the same problem of facing the lack of the space that would be necessary for waste material disposal.

It is estimated that 13.2 million tons of waste material per year (or 2.97 tons per resident) is produced in Croatia. At the same time, the need for natural materials for civil engineering projects is growing apace.

To summarize, the comprehensive utilization of different waste materials in civil engineering will have significant benefits, including a substantial reduction of dumping and stockpiling of waste, the protection of natural resources and reduction of energy requirements associated with the obtaining of natural materials, and finally, the possibility of altering or modifying properties of these basic materials to produce special engineering devices and tools for specific applications [1].

The steel industry is one of the biggest polluters of the atmosphere, as the emission of carbon dioxide (CO_2) is proportional to the amount of steel produced. In addition, this industry creates large amounts of waste materials/by-products that must be appropriately disposed of. In order to reduce the negative impact of the steel industry on the

Characteristics and Uses of Steel Slag in Building Construction. http://dx.doi.org/10.1016/B978-0-08-100368-8.00008-7

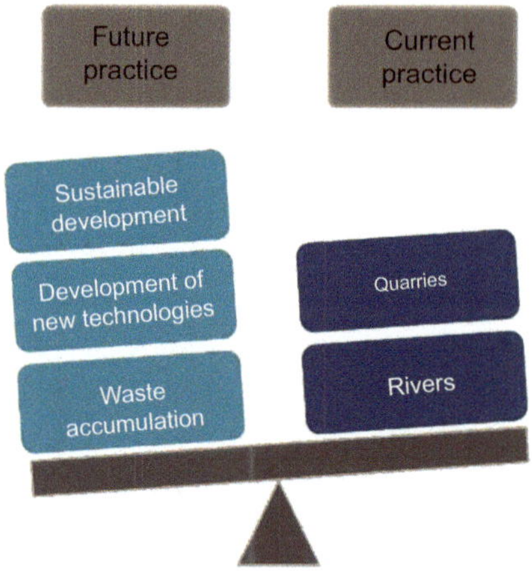

Figure 8.1 Current and future civil engineering practices

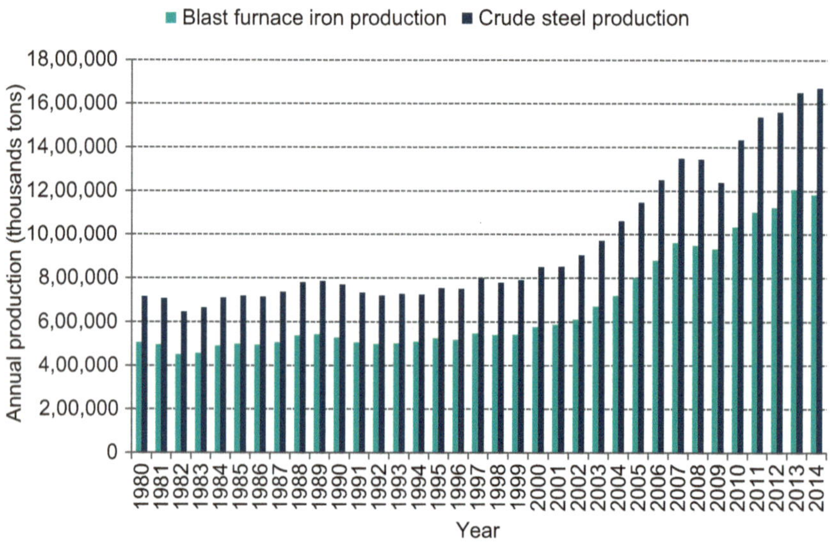

Figure 8.2 Annual blast furnace iron and crude steel production in the world from 1980 to 2014

environment, numerous studies have been conducted to explore new ways of recycling slag as a by-product in steel production.

As stated in previous chapters, slag utilization has great potential in civil engineering. But for the sake of assessing the real possibilities for slag application in civil engineering, it is necessary to provide an overview of the available quantity of this material. Figure 8.2 shows the amount of blast furnace iron and crude steel produced in the world between 1980 and 2014 according to recent data from the World Steel Association official website [2, 3], and it is given here with the aim to estimate the total quantity of available slag.

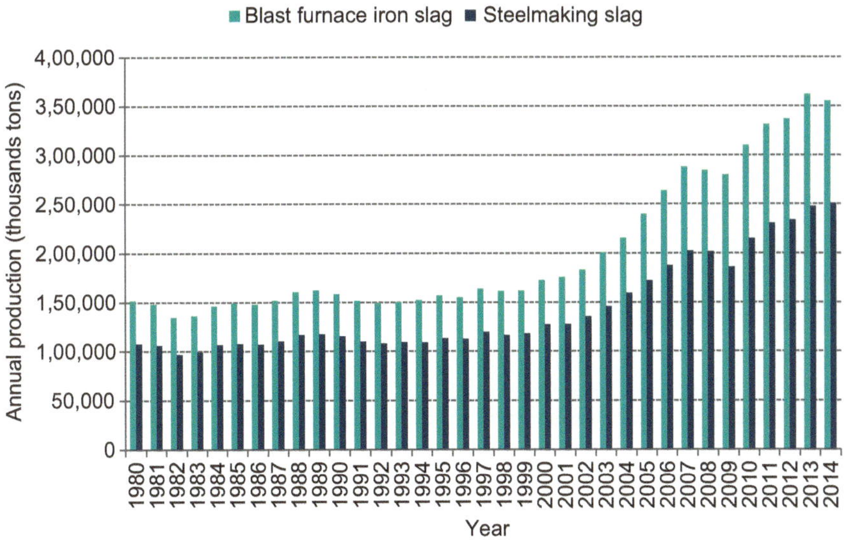

Figure 8.3 Estimated amount of blast furnace iron and steelmaking slag generated in the world each year from 1980 to 2014

The amount of slag produced is largely determined by the overall chemistry of the raw material charges in the furnaces [4]. Therefore, estimating the total amount of slag in the world that is generated in a given period requires some assumptions on the amount of slag developed per tonne of iron or steel.

The data available in the research literature vary. According to [4], for a blast furnace, the chief determinant that influences the slag quantity is the overall grade of the iron ore. For an ore feed grading 60%–66% iron, 0.25–0.30 tonnes of blast furnace slag will be produced per tonne of crude iron. Lower-grade ores yield more slag—sometimes as much as 1.0–1.2 tonnes of slag per tonne of crude iron. On the other hand, the quantity of 0.25–0.3 tonne of slag per tonne of iron is recorded in [5]. Steel slag output also varies, depending on both the feed chemistry and the type of furnace used, but is typically about 0.2 tonne of slag per tonne of crude steel [4]. According to [6], the quantity of slag developed per tonne of steel produced in the electric arc furnace is 0.15. According to [7], Nippon steel specifies 0.3 tonne per tonne of pig iron for blast furnace slag and about 0.1–0.15 tonne per tonne of molten steel for steelmaking slag. To create Figure 8.3, which shows the total estimated amount of ferrous slag in the world generated in the period 1980–2014, the authors adopted the assumption of 0.3 tonne of iron slag per tonne of pig iron and 0.15 of steel slag per tonne of crude steel, as those were the most frequently mentioned data in the literature.

Not taking into account slag generated before 1980, which may have been used already in some civil engineering structure or lies trapped in a forgotten landfill somewhere, it can be concluded that a huge amount of potentially useful material is available, which could replace natural resources. In order to handle such a huge amount of slag, slag associations in some countries, as well as in Europe, were established that focus on the manufacturing and uses of slag and the promotion of slag as a product. Some of these associations are listed in Table 8.1.

Table 8.1 Associations related to slag

Country/Region	Name of association	Year founded
United States	National Slag Association	1918
Germany	Technical Association for Ferrous Slag (Forchungsgemein-schaft Eisenhuetten-schlacke)	1968
Japan	Nippon Slag Association (NSA)	1978
United Kingdom	Cementitious Slag Makers Association	1985
Australia	Australasian (Iron and Steel) Slag Association	1990
Canada	Canadian Slag Association	–
Europe	European Slag Association	2000
United States	Slag Cement Association	2012
Brazil	Brazilian Slag Association	–

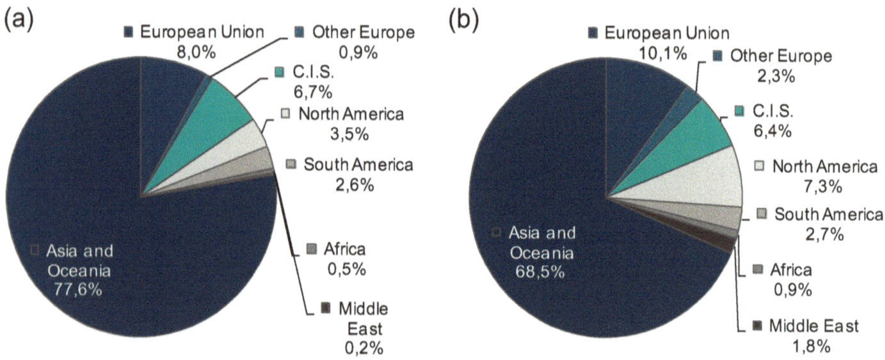

European Union: Austria, Belgium, Bulgaria, Croatia, Czech Republic, Germany, Finland, France, Greece, Hungary, Italy, Latvia, Luxembourg, Netherlands, Poland, Portugal, Romania, Slovak Republic, Slovenia, Spain, Sweden and United Kingdom; Other Europe: Albania, Bosnia-Herzegovina, Macedonia, Montenegro, Norway, Serbia, Switzerland and Turkey; C.I.S.: Azerbaijan, Byelorussia, Kazakhstan, Moldova, Russia, Ukraine, Uzbekistan. North America is Canada, Cuba, El Salvador, Guatemala, Mexico, Trinidad and Tobago, United States. South America: Argentina, Brazil, Chile, Colombia, Ecuador, Paraguay, Peru, Uruguay, Venezuela; Africa: Algeria, D.R. Congo, Egypt, Ghana, Kenya, Libya, Mauritania, Morocco, Nigeria, South Africa, Tunisia, Uganda, Zimbabwe; Middle East: Iran, Israel, Jordan, Oman, Qatar, Saudi Arabia, Syria, United Arab Emirates. Asia: China, India, Indonesia, Japan, D.P.R. Korea, South Korea, Malaysia, Mongolia, Myanmar, Pakistan, Philippines, Singapore, Sri Lanka, Taiwan, Thailand, Viet Nam; Oceania: Australia, New Zealand

Figure 8.4 Share of various regions in the total iron (a) and steel production (b) in 2014

To exchange information concerning the management of metallurgical slags in their countries, Australasian (Iron and Steel) Slag Association, Brazilian Slag Association/Brazil Steel Institute, Canadian Slag Association, Euroslag, National Slag Association, and Nippon Slag Association have joined to form a network called the World of Iron and Steel Slag.

The authors of this book found it interesting to take a look at the distribution of certain type of slag by regions of the world. Figure 8.4 shows the share of each region in the total iron and steel production in 2014 according to [2, 3] and thus the distribution of the total amount of blast furnace/steel slag generated by each region. Asia and Oceania are the regions where the greatest quantity of both types of slag is generated. In all other regions covered by Figure 8.4 both slag types are equally covered.

While all iron products made today come from blast furnaces, steel is mainly produced in basic oxygen furnaces (BOFs)—73,4% of the world production, in fact [3]. Steel production in Siemens-Martin (SM) furnaces is almost nonexistent today, being only a negligible percentage (0.6% of the world's production) in Asia. Using data on crude steel production by the process/furnace used [3], an assessment was made on the share of each steel slag type in the total steel slag production in Europe for 2013 (see Figure 8.5). "Europe" here refers to the European Union plus Other Europe. It is evident from Figure 8.5 that most of Europe uses the basic oxygen process in the production of steel.

While blast furnace slag has a long history of use as an industrial by-product, going back almost 100 years in the United States and 150 years in Europe [8], steel slag was less popular. This can perhaps be explained by the vast amount of blast furnace slag available [6]. However, this statement can be considered only partially correct; although blast furnace slag is available in the world in great amounts, its quantity today is not so dominant over steelmaking slag. Therefore, both slag types are interesting to scientists today, who are looking into the possibility of developing new uses for them. An overview of the possible uses of iron and steelmaking slag in civil engineering (given in Chapters 4 and 5 in greater detail) is summarized in Table 8.2.

While slag usage in civil engineering has been shown to be justified on a scientific level, its practical application has failed in some countries. Figure 8.6 shows a comparison of the amount of the steelmaking slag generated per capita in the period 1980–2014 [3] in Germany and Japan, as examples of countries in which slag has been incorporated into practical applications, as opposed to Croatia.

The failure in the practical application of steel slag in Croatia is probably due to the small amount in which it is formed and the variations in its produced quantities over the analysed period. All of the abovementioned data result in inconsistent quality of the slag, so it is difficult to standardize and find permanent areas where it can be used. In contrast, Germany and Japan have a long tradition of producing and using steel slag. Moreover, Japan and Germany have standards for its application, whereas Croatia does not.

In Europe, there is a tradition of slag utilization standardization, but only in countries with a longstanding connection to the steel industry. For example, in 1909, the first German standard for Portland slag cement (slag contents < 30%) was published, and in 1917, the first standard for blast furnace cement (slag contents < 85 %) was published [9].

In 1941, the German norm DIN 4301 was published, which described the necessary properties of steel slag and setting standards to comply with [10]. The use of steel slag is regulated further via the specifications that deal with the quality of mineral aggregate for asphalt manufacturing. In Germany, the document "Technical Terms

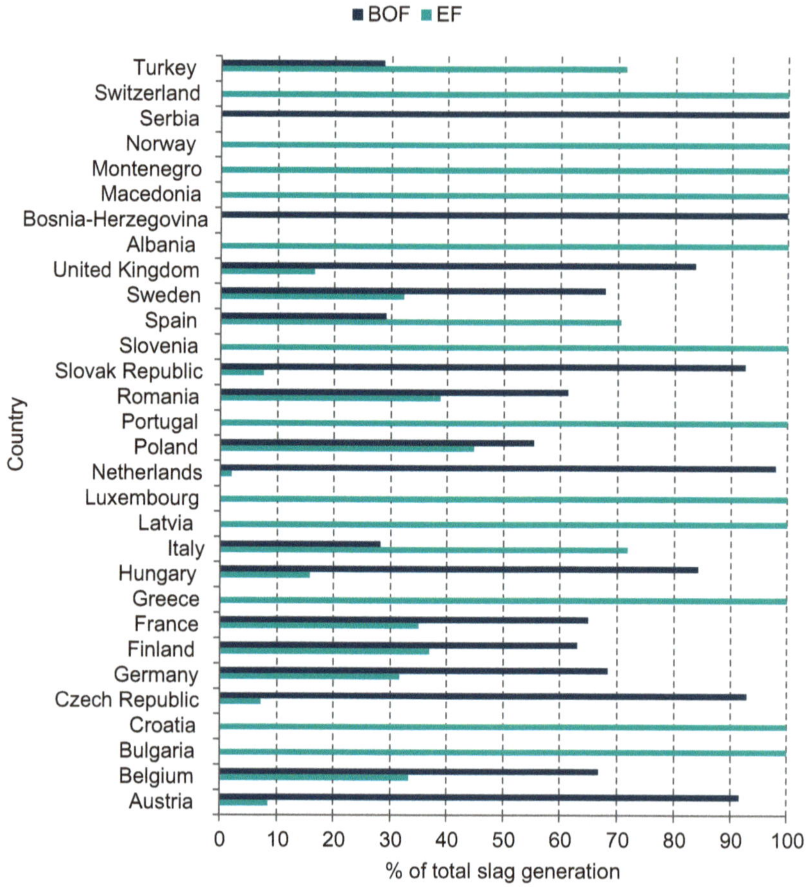

Figure 8.5 Steel slag generation in Europe by process/furnace used in 2013

Table 8.2 An overview of the possible uses of iron and steelmaking slag in civil engineering

	Blast furnace slag	Steelmaking slag
Addition in Portland cement production	x	
Mineral admixture for concrete preparation	x	
Kiln feed in cement clinker production		x
Independent binder in mortar/concrete	x	
Independent binder in soil stabilisation	x	
Aggregate in concrete	x	x
Aggregate in mortar	x	x
Aggregate in pavement unbound base layers	x	x
Aggregate in cement stabilized base courses	x	x
Aggregate in asphalt mixes	x	x
Aggregate in concrete pavements		x
Armourstone in hydrotechnical structures		x
Aggregate for acid mine drainage treatment		x

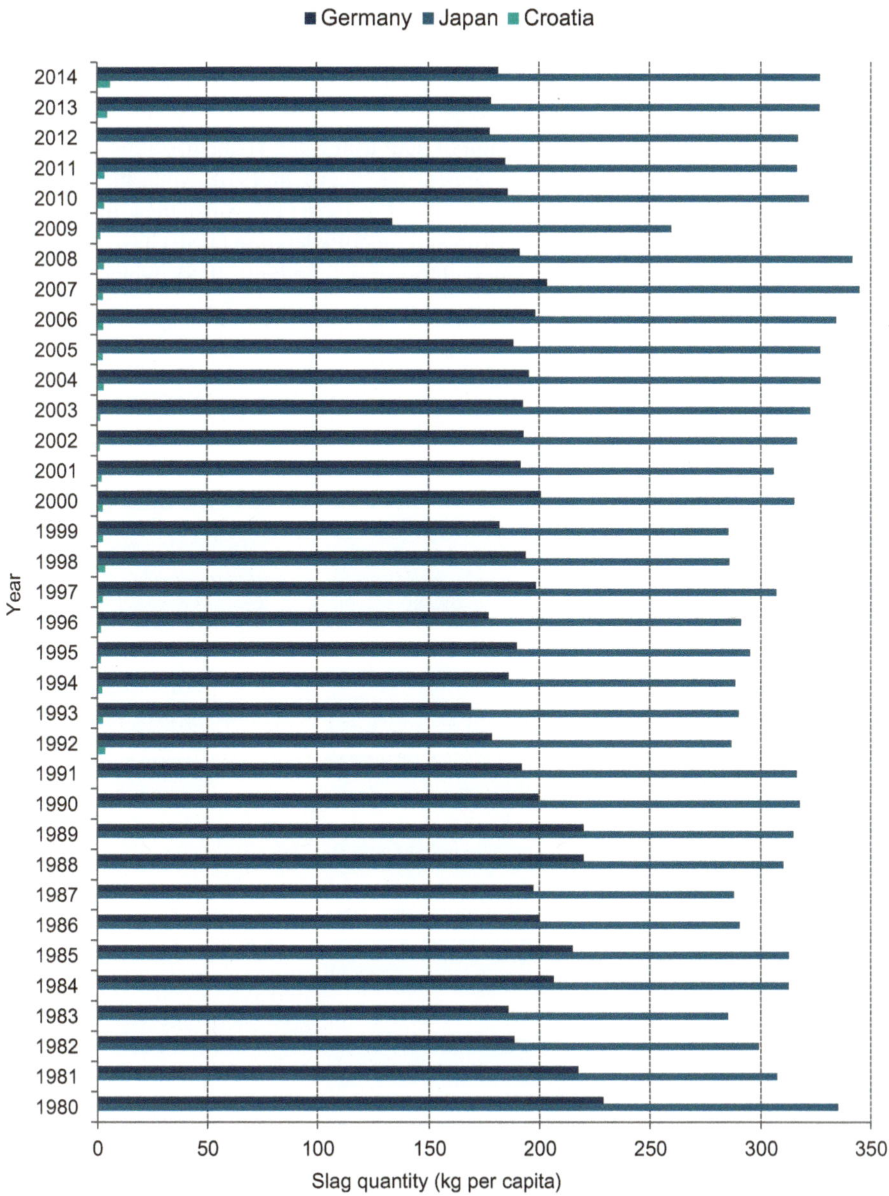

Figure 8.6 Comparison of steel slag generated from 1980–2014 for Germany, Japan, and Croatia

of Delivery for aggregates used in road construction, published in 2004, amended in 2007" has laid out the relevant test methods and norms for slag. This document regulates the frequency of mandatory testing and certification by independent laboratories and describes the criteria for internal quality assurance that the steel works which bring the slag to market must meet.

In the United Kingdom, the first standard for Portland slag cement was published in 1923, while in 1986 the first British standard for ground granulated blast furnace slag as a concrete addition was released [9]. In Chapters 3, 4 and 6, more details about current European standards were discussed.

In Japan, there is also a long tradition of slag usage and standardization. The production of Portland blast furnace slag cement has begun in 1910, while the Japanese national standard for that type of cement was set in 1926 [11]. Since then, the Nippon slag association (NSA) and the Japan Iron and Steel Federation (JISF) have promoted the institution and adoption of Japanese Industrial Standards (JIS), which are presented in Figure 8.7.

Today, the EU Waste Framework dictates the goal for various types of waste to be recycled, reused, or otherwise recovered by 2020. To keep up with those demands, many universities, institutes, and government organizations have been involved in the

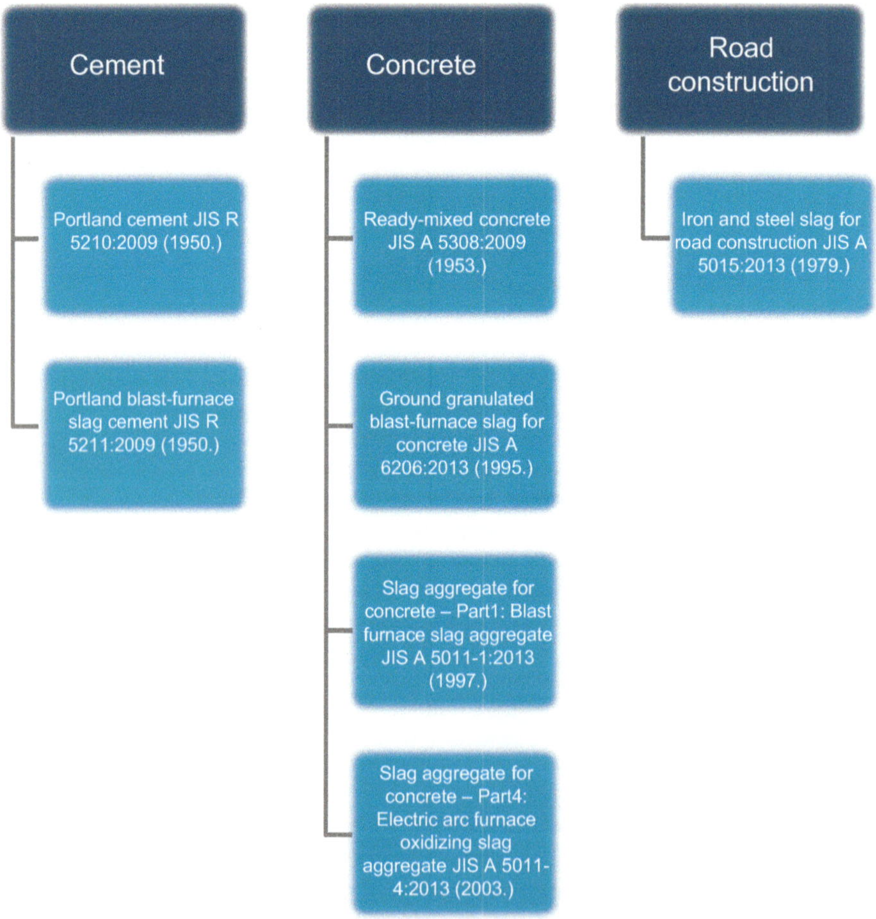

Figure 8.7 Japanese Industrial Standards of iron and steel slag [12]

research and development of secondary or recycled materials. Some of the first projects supported by the European Commission dealing with waste reuse were ALT-MAT (ALTernative MATerials in Road Construction) and COURAGE [13, 14], begun in 1998 and 1999 respectively, which gave insight into the use of these materials and helped in creating a legal framework for it.

The concepts of sustainability should be the norm for all aspects of human development today. In the construction industry, this mandates the conservation of natural resources and minimal impact on the environment. The application of secondary "alternative materials" in civil engineering, which features the most exploitation of natural materials, presents a more rational and economically and ecologically suitable solution. The term *alternative materials* covers all materials except natural virgin materials, and usually refers to construction and demolition waste and a variety of industrial by-products. In the countries of the European Union, a wide range of such materials have been used as a substitute for standard building materials for a number of years. For these materials to be applicable, they should meet certain criteria, show a satisfactory level of performance, and be cost-effective when compared to traditional materials.

Many studies have been carried out dealing with the technical characteristics of alternative material usage in civil engineering, and many of them have been presented in this book. But today, it has become increasingly important to devise new standards for eco-friendly materials and products and to investigate potential adverse environmental impact of alternative materials. In Japan, these issues are incorporated in 2011 within the "Guidelines for Introducing Environmentally Sound Qualities and Inspection Methods into the JIS of Slags for Concrete and Road (Text and Explanations)" by the Technical Committee on Civil Engineering and the Technical Committee on Architecture [12]. According to these guidelines, the most critical exposure environment of slag aggregate used for general purposes is the reuse of a demolished concrete structure as a subbase course material. Also, the most critical environment of using slag for road construction is either the condition of slag reuse as a subbase course or subgrade material when the slag was used for subbase, or the condition of slag reuse as a subbase course material after mixing in asphalt when the slag was used for a hot asphalt mix. In either case, environmentally sound qualities of the slag aggregate are specified in accordance with the applicable soil environment standards, groundwater environment standards, and specified standards (elution levels, chemical compositions).

When using waste materials as slag in civil engineering, the environment and human health can be threatened in three ways. First, some waste and by-products may contain concentrations of trace metals (such as As, Cd, Cu, Cr, Hg, Pb, and Zn) or trace organics (benzol, phenol) that are higher in concentration, more environmentally mobile, or both than those found in conventional materials. Second, it may contain highly alkaline materials (e.g., free lime), high concentrations of soluble salts, or very fine particles that may be susceptible to dusting and may also be respirable. Third, it may contain volatile organic material (such as chlorinated hydrocarbons) or inorganic material (such as As, Hg, Cd, Pb, and Zn) that could be released in high-temperature environments [15]. A detailed analysis of the potential adverse environment impact of slag usage in civil engineering is presented in Chapter 3 of this book.

In accordance with the material covered in previous chapters in this book, it can be concluded that shortcomings in the use of alternative materials primarily stem from the lack of knowledge about these materials, the lack of practical experience with them, and the lack of standards and regulations, and not from the lack of the alternative materials themselves as building materials. So the authors of this book greatly agree with the findings of Sherwood [16], who laid out the following steps that could stimulate a greater use of alternative materials:

- Develop new or supplement existing standards and technical conditions for alternative materials, which would cover their fields of use and the sought-after properties. This would avoid using the current standards for using standard materials. These new technical conditions should facilitate the use of alternative materials.
- Learn from the experience of other countries in the use of alternative materials, their standards, and technical conditions. For a long time, developed European countries have been included in programmes to research the possible applications of alternative materials.
- Develop guidelines for the use of alternative materials that would also inform all participants in construction of the possibilities of use of these materials, familiarise them with the results of ongoing research, organise seminars or conferences on the topic of the possibilities of use of alternative materials.
- Alter the perspectives of members of the construction industry concerning the use of alternative materials. The main reason for the lesser use or avoidance is the lack of an initiative for the use of nonstandard materials. There is substantial positive experiences of alternative materials use, and these experiences need to be transfered to construction industry. In order to introduce new materials and technologies into the construction industry, public need to be familiar with all possible benefits and potential problems of using new technologies.
- Administrative measures to increase the use of alternative materials are needed, such as proposing the use of specific alternative materials, introducing taxes on the use of natural aggregates, introducing taxes on the disposal of waste in landfills, and introducing requirements that all projects must include a share of waste or recycled materials and preserve high-quality materials for high grade uses.

References

[1] G. Wang, R. Thompson, Slag use in highway construction—The philosophy and technology of its utilization, Intl. J. Pave. Res. Tech. 4 (2) (2011) 97–103.

[2] Blast furnace iron production, 1980–2014, Accessed March 15, 2016, from http://www.worldsteel.org/statistics/statistics-archive/iron-archive.html.

[3] Steel Statistical Yearbook 2015, Accessed March 15, 2016, from http://www.worldsteel.org/dms/internetDocumentList/bookshop/2015/Steel-Statistical-Yearbook-2015/document/Steel%20Statistical%20Yearbook%202015.pdf.

[4] H. van Oss, Slag—Iron and Steel, Accessed March 20, 2015, from http://minerals.usgs.gov/minerals/pubs/commodity/iron_&_steel_slag/islagmyb02.pdf.

[5] Settling Drums, Design of to Slag, Iron and Steel, Supply-Demand Relationships, in: J.J. McKetta Jr. (Ed.), Encyclopedia of chemical processing and design: Volume 50, CRC Press, Taylor and Francis, New York, 1994.

[6] A. Rastovčan-Mioč, T. Sofilić, B. Mioč, Application of electric arc furnance slag, in: K. Grilec, G. Marić (Eds.), Matrib 2009. Korčula: Hrvatsko društvo za materijale i tribologiju, 2009, pp. 436–444.

[7] K. Horii, et al., Processing and reusing technologies for steelmaking slag, Nippon Steel Technical Report No. 104, August 2013, Accessed March 21, 2015, from http://www.nssmc.com/en/tech/report/nsc/pdf/104-20.pdf.
[8] Global CCS Institute, Introduction: Blast furnace slag and steel furnace slag applications. Accessed March 21, 2015, from http://hub.globalccsinstitute.com/publications/51-introduction-blast-furnace-slag-and-steel-furnace-slag-applications.
[9] EUROSLAG, Cement/concrete, Accessed March 23, 2015, from http://www.euroslag.com/applications/cementconcrete/.
[10] G. Winkelmann, U. Schaller, Steel slag: Conversion of an industrial waste material into a value adding asphalt ingredient. Accessed March 23, 2015, from http://www.aapaq.org/q/2011st/CAPSA2011/FA2_08_193_1116.pdf.
[11] T. Isawa, Update of iron and steel slag in Japan and current developments for valorisation, in: A. Malfliet et al. (Ed.), 3rd International Slag Valorisation Symposium, Leuven, Belgium, 2013, pp. 87–98.
[12] T. Sasaki, Standardization of iron and steel slag products, Nippon Steel and Sumitomo Metal Technical Report No. 1092015. Accessed March 24, 2015, from http://www.nssmc.com/en/tech/report/nssmc/pdf/109-30.pdf.
[13] Transport Research and Innovation Portal, ALT-MAT—Alternative materials in road construction, 1999, Accessed March 20, 2015, from http://www.transport-research.info/project/alternative-materials-road-construction.
[14] Transport Research and Innovation Portal. COURAGE—Construction with unbound road aggregates in Europe, 1999, Accessed March 22, 2015, from http://www.transport-research.info/project/construction-unbound-road-aggregates-europe.
[15] W. Chesner, et al., The user guidelines for waste and byproduct materials in pavement construction, FHWA-RD-97–148, Federal Highway Administration, Washington, USA, 1998.
[16] P.T. Sherwood, Alternative materials in road construction: A guide to the use of recycled and secondary aggregates, Thomas Telford Publishing, London, 2001.

Index

'*Note:* Page numbers followed by "f" indicate figures, "t" indicate tables.'

D

E

W